PERMACULTURE

PERMACULTURE

*A Student's Guide to the Theory
and Practice of Ecovillage Design*

Jan Martin Bang

Floris Books

First published in 2015 by Floris Books
© 2015 Jan Bang

Jan Bang has asserted his right under the
Copyright, Designs and Patent Act 1988 to be
identified as the Author of this work

 This book is also available
as an eBook

British Library CIP Data available
ISBN 978-178250-167-1
Printed & Bound by MBM Print SCS Ltd. Glasgow

Contents

Author's Foreword

It is now a couple of decades since the Global Ecovillage Network was launched at Findhorn, and the idea of Ecovillages has spread throughout the world. During this time the Ecovillage idea has been transformed according to the contexts in which it has found itself, so Ecovillages in Senegal look very different from Ecovillages in Northern Europe. A multitude of training programmes, seminars and courses have evolved to support people who are inspired by the idea. These have also evolved according to circumstance, context and the individuals involved.

However much these initiatives might differ from place to place, there are certain patterns and principles they all have in common.

First of all, change comes from within each individual, so here we have a common starting point. Our thinking patterns, which have created a world with existential problems of climate, pollution, economic meltdown and social conflicts leading to violence and war, need to change to patterns that will allow us to live in harmony with each other, and with our planet.

Secondly, we need a multitude of tools, technical, social and spiritual, in order to manifest this new thinking into concrete realities in our world; buildings, gardens, farms, communities and social organisms.

Thirdly we need to find partners with whom we can create larger realities. It would be a mistake to think that Permaculture or Ecovillage building is the only way to 'save the world'; both mistaken and arrogant. Our way into a sustainable future is a broad path, indeed several parallel paths, all leading in more or less the same direction. So we need to reach out to biodynamic farmers, to Community Supported Agriculture projects, to environmental builders and ecological designers, to natural doctors and healers, to all who share the new paradigm.

This book is based on my experience in teaching Permaculture and Ecovillage design in Northern Europe and the Middle East for two decades. I am struck by the wide differences between Iceland and Palestine, between Latvia and the United Kingdom, and between Norway and Israel. The book arose out of two ideas. The first was that my first book, *Ecovillages,* is now ten years old, and much water has passed under the bridge since then, certainly in my personal story, and I have learnt a great deal from teaching in new places. Secondly, I have been highly inspired by the publication of *The 4 Keys* by the Gaia Education group, a series of texts designed to be used by participants in the international Ecovillage Design Education courses. I have been teaching the Permaculture Design Course, sometimes angled for those wanting to contribute to Ecovillage building, but more often to design their way to a more sustainable lifestyle and culture. It seemed to me appropriate to bring the ideas from these books into concrete teaching and learning situations.

This is not simply an academic study. I have tried to make the subject accessible to all, and have included many bits of advice gleaned from years of teaching. Some of these appear as exercises that can be done as a group, working on the ideas presented in the various chapters. These are tried and tested exercises that I use when I teach, and have been found to be very helpful. In the evaluations we carry out at the end of all courses, I have found that these exercises score highly.

My thanks go first to the team at Floris Books, Christian Maclean, Katy Lockwood-Holmes, to Leah McDowell for her inspiring cover design and especially to Christopher Moore for his tremendous help in getting my chaotic manuscript into its final shape. It is now a decade since we started working together, this is our fifth book and I could not wish for a more supportive and cooperative team to publish with.

Thanks also to Hildur and Ross Jackson, Albert Bates, May East and Frederica Miller for conversations, support and many fruitful ideas. Out of these conversations came the idea that Kosha Joubert, a leading figure in the Global Ecovillage Network, would contribute to the book. So thank you, Kosha, for the following: the guest preface on *Ecovillages in a Time of Transition; Dragon Dreaming* with John Croft; *Ecovillage Design Mandala* and *Transition to Resilience* together with Robin Alfred at the end of the book, Appendices 1 to 3. My thanks also to many others who have helped along the way: Mike Kaplin, Cathrine Dolleris, Ann Ellsie Morger for her short account of Auroville, Laura Podoski for her photographs of Auroville, Chris Coates and Kristin Vala Ragnarsdottir.

My thanks also to my family and to the wider community of Åsbygda in Southern Norway where I live. You are providing me with a safe haven in an otherwise stormy and unpredictable world, a base for me to return to, and a fellowship that nurtures me. Special thanks also to our daughter-in-law Aline Fohre, and our daughter Sarah Bang, for permission to use photographs of Straw Bale building and from Kibbutz Lotan.

Ecovillages are offering solutions, and as such, are maintaining an honourable tradition that has been a feature of Western Civilisation for at least 2,500 years. I am proud to offer this book as a small contribution to this tradition, and hope that it will inspire and help more people to join us in exploring sustainable ways into the future.

Guest Preface
by Kosha Joubert

Ecovillages in a Time of Transition

In April 2014 the Intergovernmental Panel on Climate Change (IPCC) published the third part of a trilogy of reports compiled by thousands of the world's most eminent scientists, giving the most accurate and up to date account of climate change. The first report, released in September 2013, showed that climate change was 'unequivocally' caused by human activity. The second, published in March 2014, warned that the impact of global warming, from extreme weather patterns to reduced food production, posed a grave threat to humanity and could lead to wars and mass migration. The third draft report, published in April 2014 stated that the world must urgently switch to clean, renewable sources of energy.

Community-led responses to climate change are a necessary part of any whole system change towards resilient societies. Already in 2009, the vice-president of the IPCC, Mohan Munasinghe, speaking about the importance of the role of citizens in regard to climate protection, said: 'Change needs to come from the grassroots level, simply because the establishment moves only very slowly.'

The Global Ecovillage Network (GEN), set up in 1995, is an umbrella organisation working to support the experimental creation and preservation of low impact lifestyles across the globe. GEN serves as an alliance between rural and urban, traditional and intentional communities, and encompasses communities with some of the lowest recorded per capita carbon footprints. Through the sharing of best-practice and innovative solutions, and the honouring of deep-rooted traditional knowledge and local cultures, GEN is creating a pool of wisdom for sustainable living on a global scale. GEN supports the paradigm of thinking globally while acting locally.

GEN-International works through five broad regional organisations: the Ecovillage Network of the North Americas (ENA), the Latin American Ecovillage Network (El Consejo de Asentamientos Sustentables de las Américas / CASA), GEN Oceania and Asia (GENOA), GEN-Europe/ Middle East (GEN-Europe), and GEN-Africa. These networks make visible the impressive work done by at least ten thousand communities on the ground in more than one hundred countries worldwide.

So what is an Ecovillage?

> Ecovillages are intentional or traditional communities, consciously designed through locally-owned, participatory processes to regenerate social and natural environments. The four dimensions of sustainability (ecology, economy, social and cultural) are integrated into a holistic sustainable development model that is adapted to local requirements.

Ecovillages combine a supportive and high-quality social environment with a low-impact way of life. Rapidly gaining recognition as demonstration sites of sustainability in practice, Ecovillages are increasingly places of inspiration for the wider society that is trying to find answers to climate change and peak oil (see below, Chapter 12). They demonstrate that it is within human capacity for communities to consciously enhance and improve their social and natural environments. As such, they are precious playgrounds in which civil society engagement can come to the fore.

As the IPCC reports show, we urgently need to learn how to design societal change processes: every village and every city needs to become sustainable and needs to start regenerating their natural and social environments if we are to find a healthy response to climate change. One of the core questions GEN is facing is, therefore, how the Ecovillage approach can be disseminated at scale without losing its core value of locally owned, participatory processes?

We also know that however important the role of civil society efforts in whole-society change processes may be, for them to become truly effective and influential, they need to be closely coordinated with benevolent governmental and corporate support and collaboration.

Fortunately we can learn from current examples in Senegal and Thailand, where such bridges between civil society, governmental decision makers, and the corporate sector have been built in order to support community-led responses to climate change on regional and national levels.

In Senegal, a network of forty-five Ecovillages came into being following the delivery of educational sustainability programs (forerunners of the Ecovillage Design Education), soon after the establishment of GEN. In 2002 GEN Senegal was established to provide a forum where civil society could showcase best-practices (integration of solar cookers, drip irrigation, permaculture design, reforestation programs, etc.) and demonstrate the potential of an Ecovillage approach for sustainable development. The Senegalese government, inspired by the work done on the ground, then established the Senegalese National Agency for Ecovillages, ANEV, positioned within the Ministry of Environment and Sustainable Development. Today, Senegal is the first country in the world to create an integrated national Ecovillage transition program dedicated to transitioning 14,000 traditional villages to Ecovillages. At the time of writing, similar, more regional, programs are emerging in Thailand and possibly Burma.

In such programs, Ecovillage transition strategies are scaled up, not through replication of a standardised *model*, but through diffusion and local adaptation of the underlying *principles*. The participatory nature of such processes necessitates learning from and working closely with each of the communities involved (for the example of Kartong, Gambia, see Appendix 2 on the Ecovillage Design Mandala). Every Ecovillage will be different. Such a journey of self-empowered local change can be unleashed through an Ecovillage Design Education (or a Permaculture Design Course). A process like the Transition to Resilience journey can be used as follow-up to ensure successful implementation. A combination of such trainings functions as a support system for a local community not only to design their own pathway into the future (a 'Community Resilience Plan', see below), but also for them to aim to influence good governance and effective implementation through a bottom-up 'Adaptive Governance Cycle'.

Such a cycle starts by:

- Identifying hot-spot communities and key players in the region or country.
- Facilitating a participatory community and multi-stakeholder appraisal of the community context.
- Recognising the strengths and leverage points in each of the four dimensions (ecology, economy, social and cultural) within the community context.
- Co-creating a Community Resilience Plan with all relevant local actors.
- Working with local/district authorities to integrate this plan within wider development plans.
- Acting to implement the Community Resilience Plan.

Once this has happened, and thus an Ecovillage transition process has taken place, we can go on to:

- Draw lessons learned and disseminate best-practice solutions across regions.
- Build multi-stakeholder alliances to work with local and district authorities on Regional Resilience Plans for Communities.
- Work with national governments on formulating National Action Resilience Plans for Communities.
- Bring forward adaptive policy inputs to strengthen sustainable development strategies on a national level.

In order to support societal transition, we need to marry bottom-up with top-down processes. This Adaptive Governance Cycle enables local solutions to be woven into new tapestries for resilient societies. We come a step closer to establishing a true democracy that trusts in the inherent wisdom and goodwill of its citizens and their willingness to be part of the solution.

(This article was originally published as 'Ecovillage Transition: Scaling up community led change processes'.)

Adaptive Governance Cycle: 10 Steps to Resilient Communities

1. Identify hot-spot communities and key players

2. Facilitate participatory community & multi-stakeholder appraisal

3. Recognise strengths and leverage points in each of the four dimensions

4. Co-create Community Resilience Plan: design own pathway into the future

5. Inform local/district authorities and adapt development plans

6. Act to implement changes in behaviour, norms and infrastructure

7. Draw lessons learned and disseminate best practice solutions

8. Build multi-stakeholder alliances to create Regional Resilience Plans for Communities

9. Formulate National Action Resilience Plan for Communities

10. Bring forward adaptive policy inputs for sustainable development

*View over Dyssekilde
Ecovillage in Denmark.*

Chapter 1. Ecovillages

Human beings have always lived in community. Indeed, it is impossible to think of a human individual being able to survive without community. Who would feed and clothe the baby? Who would teach the toddler to talk? Who would care for the old folks?

For most of our existence we lived in small hunting and gathering groups, tied together by family. About ten thousand years ago we began tending animals and cultivating crops, though the transition was clearly very gradual, and not at all evenly spread throughout the world. Even today there are groups in many places throughout the world that still live by hunting and gathering.

As cultivators, gardeners, farmers, or animal herders we lived mostly in small villages, and only a few people in more recent times lived in what we might call towns or cities. The phenomenon of urban living that we see today is very recent. A tiny percentage of us lived in cities during the last five thousand years, and only during the last few centuries has the drift to urban living gathered pace. The present statistics have been achieved only over the last generation. City living is so recent that we have hardly developed the appropriate responses, socially and psychologically, to deal with it.

Throughout our long history, human communities have been dominated by tradition. That is how we have transmitted our culture and learning. Tradition is an amazing human invention. It has enabled us to build on the knowledge and experience of our forebears, giving us skills that we could not gain in just one lifetime. But of course tradition is a two-edged tool, it can also stifle new ideas and innovations.

When I first heard of what we now call intentional community, it was called communes. A fine word at the time, but during the 1960s and 70s it collected a good deal of media attention, and was linked to hippies, group sex and drugs. Actually there weren't much of these in most communes and we subsequently realised that this was the media's attempt to trivialise the whole issue. So in order to coin a term that was less loaded, we began using 'Intentional Community', a term that I think is much better anyway.

This was not, however, a new trendy fad thought up by a few weirdos in the 1960s. The idea of 'Intentional Community' was invented by courageous pioneers who wanted to innovate new ideas, and experiment with new social forms and relationships. I write courageous deliberately, because we shall see that many of these pioneers have had to suffer for their break with traditions, in many cases paying with their lives. This idea of creating experiments in social living has a long history.

Intentional communities in the distant past

Pythagoreans

The earliest mention of intentional community seems to be around the time of Pythagoras and Plato. The latter's book *The Republic* marks the beginning of a long and honourable tradition of utopian literature, and the book has had considerable influence upon social thinking throughout history.

While Plato thought and wrote, Pythagoras thought and did. He was born on the island of Samos around 570BC and moved to Croton in Italy around 530BC. On his arrival he exerted a broad influence, and many people began to follow his teachings. Later biographers mention the effects of his eloquent speech in leading the people of Croton to abandon their luxurious and corrupt way of life and devote themselves to the purer system which he came to introduce.

His followers established a select brotherhood or club for the purpose of pursuing the religious and ascetic practices developed by their master. What was done and taught among the members was kept a secret, but temperance of all kinds was a dominating feature of their community. For instance, there is disagreement among the biographers as to whether Pythagoras forbade all animal food, or only certain types. The Pythagoreans were a philosophical school, a religious brotherhood, and a political association. Such an aristocratic and exclusive club could easily have made many people in Croton jealous and hostile, and this was what led to its destruction. The circumstances are uncertain. An attack was made upon them while assembled. The building was set on fire, and many of the members perished, only the younger and more active escaping. Similar commotions ensued in the other cities of Italy in which Pythagorean groups had been formed.

As an active and organised brotherhood the Pythagorean order was suppressed, and did not again revive. Still the Pythagoreans continued to exist as a sect, the members of which kept up among themselves their religious observances and scientific pursuits, while now and then individuals acquired great political influence. This had a large effect on future esoteric traditions, such as the Rosicrucians and Freemasons, both of which were occult groups dedicated to the study of mathematics and both of which claimed to have evolved out of the Pythagorean brotherhood.

Reginald Allen, in his book, *Greek Philosophy*, credits the Pythagoreans with founding mathematics as we know it today, and ranks it as one of the greatest intellectual contributions to civilisation.

Pythagoras is said to have written the following:

We ought so to behave to one another as to avoid making enemies of our friends, and at the same time to make friends of our enemies.
Friends share all things.
As soon as laws are necessary for men, they are no longer fit for freedom.

This is from a period at the beginning of written history. There may have been attempts at creating 'alternative societies' before then, but without written documentation, how are we to know?

Essenes

Half a millennium later we find an intentional community called the Essenes out in the desert at the northern end of the Dead Sea. They were a radical Jewish group, who left the cities and villages of Judea, today's

Palestine and Israel, and set up an alternative to their mainstream society at a place called Qumran. We know little about them, but much more than we know of Pythagoras' community in Italy. The Essenes are mentioned in Josephus' *History of the Jews*, may be hinted at in the New Testament, and have left behind a collection of writings, the Dead Sea Scrolls, that point to a theology which was post Old Testament Hebrew, and has parallels in the Christian theology of the New Testament. Some theologians argue that both John the Baptist and Jesus Christ were members or were at least closely associated with the Essenes.

Philo of Alexandria (c. 20BC to AD50) gives us one of the best contemporary accounts of the Essene community. They had no private possessions, houses and clothes were communal, they ate together, shared their incomes and the ill or needy were looked after out of the common purse.

Philo estimates their number to be at least four thousand, spread over the Levant, in today's Israel, Palestine, Lebanon and Syria. They avoided living in towns, preferring small villages in order to keep a healthier lifestyle, both physically and spiritually. They were intensely religious, but avoided animal sacrifice, something which was still the practice in the Temple in Jerusalem up till its destruction in 72AD.

Some worked the land, others made their living by handicrafts, but only enough to supply their needs. Philo points out that though they were without money or riches, they were wealthy in their frugality. They had no weapons of any kind, they were scrupulously honest, and had no slaves, and no masters. Philo's description of the Essenes could fit many intentional communities, both contemporary and in the past.

Monasticism

The next appearance of this phenomenon occurs in the same region, but this time within the context of a new religion, Christianity. I haven't been able to find any direct link between the Essenes and monasticism, but there is clearly a similarity between the two, both in lifestyle and in ideals. A major difference would be that the monasteries are all single sex, while the Essene communities were composed of families.

1

Mar Saba monastery, founded about 450 AD in the Judean Desert south east of Jerusalem.

As Christianity was becoming firmly established in the Eastern Mediterranean in the first couple of centuries AD, a tradition of hermits leaving populated areas to seek solace and inspiration in the surrounding deserts started growing in Syria, Palestine and Egypt. From single hermits there arose communities of hermits, and within a relatively short time Christian monasticism became enshrined and codified with the Rule of St Benedict amongst others. The Benedictines are still with us a millennium and a half later, surely the oldest continuous intentional community tradition in western civilisation. Monasticism spread northwards to meet Celtic Christianity to form an important element in the development of western thought and culture. Celtic Christianity was suppressed by the Catholic Church a thousand years ago, but has recently attracted increasing attention and interest. Early monasticism also spread north and east to form Orthodox monasticism, which is experiencing a strong revival today after the fall of the Soviet Union.

The recent past

As the Enlightenment laid the roots of modern thinking about science and society, we can discern the emergence of secular intentional community outside the Church and as a reaction to social change. Gerard Winstanley established the Diggers on St George's Hill south west of London in 1649. Even though the commune lasted only a year or two before being suppressed and disbanded, they created a tradition which the British Labour and Cooperative movement still regards as their beginnings, and the British Commune Movement continues to publish books and directories of Communes under the title 'Diggers and Dreamers' consciously paying homage to Winstanley and his fellow communards.

With the coming of industrialisation, social conditions worsened, and the reaction was not slow in coming. Robert Owen set up a model industrial community at New Lanark on the Clyde in 1786 together with David Dale, and later went on to establish New Harmony in the United States. With this he found his way into the history of modern communal thinking and doing. The New Lanark experiment inspired a group of people in Rochdale, Lancashire, to set up a cooperative, known as the Rochdale Pioneers, the seed of the Cooperative movement, today one of the biggest alternatives to capitalism in the world.

View over New Lanark in 2014.

Pat Conaty, a research Associate with the Cooperatives UK, writing in *Resurgence* magazine in January 2014, estimates that the global cooperative movement has about one billion members, and that the Co-op provides weekly services to about three billion people. Just in Britain, between 2008 and 2013 membership of cooperatives has increased by 20% to 13.5 million people.

New Lanark was finally closed as an industrial plant in the 1950s, but re-emerged a couple of decades later as a conference centre, used today by such communal groups as the Camphill Community, the Iona Community and the International Communal Studies Association amongst others.

The nineteenth century saw an explosion of alternative communities, in Europe, in North America, and in Australia. This was a time of unsurpassed social experimentation, with equality, women's rights, group marriages, spirituality and new economic forms being tried out. Some were found to be unworkable, and countless intentional communities foundered after a short time, but many managed to find a stable form and a few survived for generations. This period was to create a foundation for the study and development of communal living up to the present day.

Since the Second World War, especially after the 1960s, the creation of a communal counter culture has been portrayed by the mainstream media as a passing curiosity, a fad, a fashion that would soon pass. To be fair, to the innocent and ignorant bystander, it does seem as if many, if not most, of the alternative intentional communal projects either fail to get off the ground, or fall apart after a short time. Diane Leafe Christian maintains that only one in ten manages to survive babyhood for more than a couple of years. When confronted with that fact it would seem that this is indeed a doomed experiment, but on reflection, the same percentage holds true for businesses. Only about one in ten small new businesses survive to establish themselves, but no one questions that business is here to stay. In the natural world the failure rate is much higher, reaching astronomical proportions when considering how many seeds a tree might produce in the course of a lifetime compared to how many seedlings sprout from these seeds to grow into new trees.

Looking at the proliferation of communal networks, the Fellowship for Intentional Community, Federation of Egalitarian Communities,

International Communal Studies Association, Diggers and Dreamers, Camphill Communities, Catholic Worker, L'Arche, and so on, and so on, there is no doubt that this is not a passing fad, but a social phenomenon that is firmly established in our western culture. And that is exactly my premise in this short excursion into history.

One pattern that emerges clearly is that every time there is a proliferation of intentional community experiments, this coincides with a period of change and turmoil in the mainstream culture. The Pythagoreans experimented with the new lifestyle at a time when traditional Greek society was being questioned by the early philosophers. The Essenes arose as the Old Testament Jewish theology was being replaced by Christianity and by Rabbinical Judaism, and during a time of political conflict between the various factions within the Roman Empire. Winstanley invited his supporters to St George's Hill during the Civil War in England, and Robert Owen's experiments coincided with the Industrial Revolution.

It is really difficult to estimate how widespread intentional community is today. Definitions are difficult to agree upon, many communities don't want to be exposed to publicity of any sort, and others don't have the capacity to answer questionnaires.

A European group has been publishing a directory they call Eurotopia every few years. In their 2000 edition they record having sent out 1,517 letters to communities they knew about, and the directory lists 326. This shows that statistics are just the visible tip of the iceberg.

In 2009 Ralf Gering came up with some exact figures for communities worldwide. It was certainly the most comprehensive listing I have ever seen, and his definitions are clear and well thought out. Worldwide he listed 281 organisations or movements, with 3,802 actual communities, comprising 407,250 residents.

The US-based Fellowship for Intentional Community has published directories of communities every few years since 1990. In their sixth edition, the number of communities listed in North America had risen from 304 in 1990 to 1,055 in 2010. They figure that about 100,000 people live in some form of intentional community in the United States.

Despite the difficulties of reliable statistics, there is no doubt that there are more and more intentional communities being established, and that they are attracting increasing interest among researchers. Increasing numbers of intentional communities are defining themselves as Ecovillages, and it's clear that the common intention is to experiment with social, technical and ecological strategies to create sustainable futures.

Today mainstream western culture is confronted by existential systemic problems and challenges; economic collapse, environmental degradation, climate change and peak oil. Economics is challenged by a series of financial crises that seem to be unsolvable, and attempts at solutions have turned out to be totally ludicrous, as if putting large segments of the workforce in Greece and Spain out of work can help build up wealth in those countries. Our social systems are reeling from how to deal with violence and the influx of unwanted refugees and immigrants. The world has never before experienced such a large number of people who are starving, either proportionally or totally. And the main health hazard in the developed countries is obesity.

What kinds of solutions are offered to us from our mainstream institutions, from universities, from government research departments, from business, from mainstream international organisations? More of the same: bail out the failing banks that created financial breakdown; shore up the agricultural system and its economic structure and tell people it's feeding the world; build more motorways for cars and trucks using the internal combustion engine; build more airports; pass some more laws.

Ecovillages today

Ecovillage design is a new discipline which responds to the needs of the future for truly sustainable, low-carbon-footprint habitats. It has grown and developed and been defined from below, without any help from governments; truly a grassroots movement. It is built on experience and knowledge of what people want and what is possible. To become a mainstream phenomenon will require dissemination of the existing knowledge base through education and popular articles, films and books that can inspire people everywhere to get involved in building a sustainable lifestyle for all of us. The Ecovillage Design Education (EDE) is one such programme that has been offered one hundred times in 28 countries on five continents in the period 2005–2012. EDE has published four books of background reading.

This initiative was launched at Findhorn in October 2005 by a group of Ecovillage educators calling themselves The Global Ecovillage Educators for Sustainable Earth (GEESE). The key programme is the EDE, the UNITAR-endorsed four-week holistic introduction to designing sustainable settlements, based on the 'living and learning' principle. This programme draws upon the experience developed in a network of some of the most successful Ecovillages and intentional communities across the world. In four weeks participants get an overview of all you need to know to design sustainable settlements anywhere in the world. The programme has since been taught to more than 3,000 persons at a number of major Ecovillages and other educational institutions in over thirty countries and is growing steadily.

The Global Ecovillage Network was created by a group of Permaculture teachers to create situations where a multitude of Permaculture solutions could be implemented within a community, each solution linked in and supporting other solutions. The idea was to check out if this way of thinking really can work, and by doing so create educational experiments that can reach out to others, helping them to find solutions. According to their website, in the autumn of 2013 there were 670 Ecovillages in 86 countries.

In 1992 Diane and Robert Gilman defined an Ecovillage. It should:

- Be human scale, usually thought of as somewhere between fifty and five hundred members, but with exceptions.
- Be a full-featured settlement, in which the major functions of life, food provision; manufacture; leisure; social life; and commerce, are all present in balanced proportions.
- Have human activities harmlessly integrated into the natural world.
- Be supportive of healthy human development. A balanced and integrated approach to fulfilling human needs – physical, emotional, mental and spiritual.
- Successfully be able to continue into the indefinite future.

In medieval Europe villages developed naturally around a rural manor house, which was the sole governing body. Over time, the manor house would be surrounded by increasing numbers of other buildings, dwellings, and enterprises: the miller, blacksmith, herbalist, midwife, wheelwright, harness maker, livery stable owner, tavern keeper, and so on. Soon village residents outnumbered those in the manor house. The manor house was the largest single entity in the village, the oldest, and it provided the 'seed' around which the village grew. But it was no longer the sole governing body. Intentional communities function similarly. They are like catalysts or seeds that begin as 'centres of research, demonstration, and training' and over time can develop into true Ecovillages.

Passive solar warming in a row of houses in Dyssekilde Ecovillage in Denmark.

Recent studies examined the consumption patterns of Ecovillage Ithaca in the USA which comprises one hundred adults with sixty children living in sixty houses on 71 hectares. Three independent studies by Massachusetts Institute of Technology, Cornell University and Ithaca

College showed that they were using 40% fewer resources than the average US neighbourhood of the same size.

The studies showed the following details:

- 71% lower water use
- 45% work at home in the Ecovillage
- 75% of their waste is recycled and/or composted
- 90% of the surface is used for cultivation, forest, water or grazing
- 40 to 50% less energy use in their domestic buildings

Findhorn in Scotland has four hundred residents living in 181 houses. Their global hectare footprint is 2.71 gha/person, compared with a UK/Scotland average of 5.40, in other words, half the national average. Findhorn exports more renewable energy than it consumes, has its own biological water treatment system, runs a local eco-currency and most residents live and work within the community.

Similar results were recorded by Kaj Hansen writing in the Danish Ecovillage magazine *Løsnet* in December 2009. Studies in various Danish Ecovillages showed energy use and pollution to be way under the national average.

As intentional communities explore new ways of creating sustainability, they are attracting the attention of local authorities also looking for this, but unable to get very far with the pathetic solutions offered by mainstream institutions. In Moray, Scotland, where Findhorn has already celebrated its fiftieth birthday, the regional council estimates a £4 million turnover can be attributed to the effect of the Findhorn Foundation. Here in Norway, the Ecovillage at Hurdal has been in existence for over a decade, and the local municipality is now looking into establishing their region as a 'Green Incubator', sharing many of the aims that a classic Ecovillage might aspire to (see further below).

Hurdal: Celebrating Open Day in 2006.

Community profile

Hurdal Ecovillage

In the late 1990s an Ecovillage group called Kilden (The Source) was established in Norway, inspired by visits to Findhorn. They had high ambitions, and when I joined them in about 2001 there were three possible sites. Within a couple of years two of them proved impractical, while the project at Hurdal, about forty minutes drive north of Oslo, eventually became a reality.

In the beginning the project consisted of a small group who wanted to build their own houses using local materials. They struggled for years, not really growing, but establishing a good working relationship with the local municipality. In 2006 there was a day seminar on Ecovillages in Oslo, with participation from several national organisations including the Norwegian ethical bank, Cultura, the Home Loan organisation (Husbanken), researchers and project managers. The then mayor of Hurdal Municipality, Jorunn Glosli, gave a highly positive account of the municipality's relationship with the Ecovillage, ending her presentation by calling for every one of the 415 local municipalities in Norway to get their own Ecovillage!

The breakthrough for Hurdal happened when they finally created a good working relationship with a broader range of colleagues; a leading eco architect, a highly skilled and experienced Permaculture teacher and town planner, the ethical bank Cultura, the Home Loan Association, contractors and financiers. A new concept in ecological housing was developed, prefabricated out of environmentally-friendly materials and with extremely low energy usage. An extensive zoning plan for several clusters was finally accepted by county and municipality planners, and construction began during 2013. By the autumn of 2014 several families were already living on site, and the first neighbourhood was nearing completion.

The most exciting part of Hurdal's story is still unfolding. The Hurdal municipality has accepted the challenge of being one of Norway's leading green municipalities, and is actively introducing environmental standards throughout its area. The regional plan is to establish the valley of Hurdal as a 'Green Incubator', with its easy access to Oslo and to the main international airport. They are taking their model from 'Silicon Valley' in California, but instead of going the route of digital information, they are looking towards a sustainable future, and all that this will entail: housing, transport, food supply, technology, energy and social structures. Their aim is to be a flagship project, and a model that can be replicated across Norway.

For more information see: http://hurdalecovillage.no

Starting Group Work

Tips for facilitators and groups

I wrote this book after teaching Permaculture and Ecovillage design for many years. Before that I had worked in environmental projects, and before that I had been a schoolteacher. Like many skills, the best way to learn to teach is to teach, and I hope that some of my experience can be shared with others. The following is a collection of tips that would help a group to establish a learning environment. In the old days, we thought that teachers were highly informed people and that their job was to pass on this information to empty minds. Today we realise that we know all that we need to know, but we don't know that we know it. The job of the teacher is more that of a facilitator, reminding the participants of what they know, and helping them to recognise it and give it a name and a formulation. In the spirit of this realisation, I try to use the word facilitator rather than teacher or instructor.

Hopi Poem

This is a good poem to begin the course with. I often read it again at the end, and ask the group if they feel differently towards it.

> You have been telling the people that this is the Eleventh Hour.
> Now you must go back and tell the people that this is the Hour.
> And there are things to be considered:
>
> Where are you living?
> What are you doing?
> What are your relationships?
> Where is your water?
> Know your garden.
> It is time to speak your Truth.
> Create your community.
> The time of the lone wolf is over. Gather yourselves!
> We are the ones we've been waiting for.
> (Words spoken by the Elders of Oraibi, Arizona Hopi Nation)

Some suggestions for groups

This book may form the programme of a self-study group, or a course facilitated by someone inspired by the book. In any case, the first thing you would want to do is to get to know one another. I have two suggestions:

Introduction session (simple)

You might want to write this on the board or flip chart:
> *There are no strangers here, only friends who have not yet met.*
> (W.B. Yeats)

Go round the circle, asking each person to say their name loudly and clearly, where they are from, how they came to be on the course, and what they hope to get out of it. The facilitator might want to make a list, especially of their expectations of the course, and refer to this when planning material for the sessions, thus making sure that the participants' expectations are realised.

I also find it useful to say their names often, when asking questions or referring to them. This is very reassuring for the participants, they feel seen and heard, and it helps create a trusting atmosphere.

Introduction session (with pictures)

The facilitator should have built up a collection of pictures. Mine is culled from old calendars, a collection of postcards, and pictures I have cut out of magazines. Have many!

Sitting in a circle gives everyone maximum participating location.

Spread the pictures on the floor, and sit comfortably in a circle. Explain to the group that each person should select one picture that appeals, and that they will be expected to share the reason for this with the rest of the group. The introduction round would then go like this: each person says their name, where they are from, and why they selected this picture.

Again, I try to note their names, and a couple of key words, and use this to refer to them, creating that trusting, friendly atmosphere in the group.

Getting to know your names

I am not in favour of name tags. They have two drawbacks. One is that you always look at their name tag, not in their eyes. The second is that you don't generally make an effort to learn names. Why bother, you can look at the name tag!

So here is a three-step game that can be played every morning, and really, anytime during the day when energy levels drop and there is need for a break with movement and attention.

I have a little rubber ball, which looks a bit like planet Earth, so I explain we shall be tossing the ball from one to another, and please don't drop the Earth, we're here to look after it (always raises a laugh!). The game has three levels:

Level 1. Throw the ball to another person, saying your name loudly and clearly: 'My name is Jan.'

Level 2. More complicated. Throw the ball to someone, saying your name and then theirs, loudly and clearly: 'My name is Jan, and I throw the ball to Chris.'

Level 3. This is challenging. Throw the ball to someone, saying your name and theirs, loudly and clearly, and then ask them to throw it to another named person: 'My name is Jan, I throw the ball to Chris, and I ask Chris to throw the ball to Judy.'

Don't forget to tell them that forgetting a name is not shameful, and that it's perfectly all right to say 'I'm sorry, I forgot your name, can you remind me?'

Within a couple of days, everyone will know everyone's name.

What to write on?

I find that I am using complicated technology less and less as I work with groups. But I do use a board or flipchart a lot. Here there are choices and preferences. A flip chart is very good, because you can tear off the paper and tape it on a wall somewhere, and you have a record of that particular session for all to see for the rest of the seminar. You can buy flipcharts and paper in most office suppliers, but I found it much cheaper to buy a large roll of white paper, and tear off sheets as I need them.

A flip chart page with notes on can be further developed by a small group.

Flip chart being used to present a garden design.

Whiteboards are fine, but I never really trust the manufacturers of the felt tips that they don't contain terrible chemicals, and of course there's always someone who uses permanent ink felt tips eventually. Chalk boards are fine too, always reminds me of my first years as a schoolteacher way back in the 1970s. I am a little concerned that the chalk dust might not be so good for us. Both of these boards have the disadvantage that when you rub them out, that's what happens, your stuff gets rubbed out. So that's why I prefer flip charts.

A Permaculture song

Get everyone to stand up. That gets the energy going when they have been sitting, taking notes and listening for a while. The following song is sung to the tune of 'Frére Jacques', anyone can follow it:

> *What is Permaculture?*
> *What is Permaculture?*
> *I don't know*
> *I don't know*
> *Some say that we're crazy*
> *Others that we're lazy*
> *Seeds we sow*
> *Trees we grow.*

If you are working in a foreign country, or have participants who do not have English as their main language, you can ask them to translate the song into another language. It's fun, and they'll remember it.

Crossed arm exercise

This is a great way to make the participants feel a change in consciousness. Ask them all to cross their arms, and to note which arm is over which (most people put right over left). Then ask them to switch arms. There will always be some who find it difficult, and more that find it awkward or uncomfortable. Crossing your arms is just a habit, and when we change our habits it's often uncomfortable.

It's the same with thinking. We have habituated ways of thinking about the world, ways of dealing with the world. Our aim in this course is to be aware of our thought habits, and change them when we see that it is useful or necessary.

Fruit salad

When working with groups the energy levels in the room go up and down. There is nothing wrong with that, but as group facilitator, your responsibility is to keep a good group atmosphere. Opening the window for fresh air is an obvious response, but from time to time a more active approach is needed. One of the easiest and quickest is to agree beforehand that at a given signal, we generally suggest that someone says 'Fruit Salad!'. When that is said everyone has to get up, shuffle around the room and find a new place to sit. The whole thing usually takes between one and two minutes, and everyone has then woken up. This gives power to each individual who feels overwhelmed by tiredness. It is also a signal to you, the facilitator, that you need to change activities pretty soon.

Cosmic circle

Get the whole group to stand in a circle, quite close together. Ask them to shut their eyes and balance themselves with straight backs, head held up and imagining a straight line from between their feet on the floor to the top of their heads. Ask them to imagine there is a column of light within the circle, shining up into the cosmos, and down into the depths of the planet.

Now ask them to rock back ever so slightly, keeping their feet firmly on the floor. Ask them to envisage the column of light opening up like an inverse cone above them. Now ask them to rock forward very slowly, putting their weight on the front of their feet, and envisaging the cone of light focussing into a point far above their heads.

Gently and slowly, ask them to rock backwards and forwards in unison, envisaging the column of light. Slowly bring them back, asking them to open their eyes and relax and move about.

This is a good exercise to set a contemplative tone, and bring the whole group together focussing on larger issues. I often use it to set the tone for a Vision Workshop.

1

Vision workshop

We work with a process that can be described as: Thoughts – Feelings – Actions. You will have to use the collection of pictures that you used earlier for the introductions.

Developing a design as a small group.

The task is to focus on a vision that starts with the individual, using thoughts, extend this to another person, where the feelings that may be associated with explaining the vision come into play. The final step is to come to an action plan, an agreed plan to do something coordinated together with the whole group.

You might want to spend a few minutes talking about what that vision could be. We have done it with the Norwegian Permaculture Association, what vision do we have for our organisation? We may define our question by asking: What do we want to achieve as a group? This is an ideal opportunity to define a vision for an Ecovillage.

The first part of the exercise involves spreading out the pictures on the floor, and giving the group a short time so each person can pick out a picture that says something that is related to this workshop at this time. Then go round the group, inviting each person to explain how the picture relates to the vision he or she has thought of. Divide the group up into pairs, or at most three people in a small group, and invite them to talk through how their individual visions can be linked together. At this point feelings will come up, sometimes positive ones, relating to how easily the visions fit together, but there is also the possibility of negative ones, when a couple finds it difficult to link their visions together, or some other interpersonal exchange that is not so easy. It's important to talk these things through until every small group has somehow formed a common vision.

The final exercise is to come together in a plenum, and each small group outlines their common vision, and the whole group tries to find a way of linking all the visions together. This may take a while, and it's important to have a good facilitator who can link everything together in an action plan. Have one or two people writing up key words on a flip chart, making sure that each keyword accurately describes what comes up.

If your course is spread out over several days, this can be pinned up on a wall, and you can ask the participants to write sentences using the words, in order to get a number of vision statements. These can then be discussed towards the end of the course.

In this way a group can move from individual personal ideals to a group plan to implement an idea in the world. It can be a powerful tool for designing a course of action for an Ecovillage, often spreading the various tasks out amongst those who are most committed to a small part of the total whole.

The ball game

It's the afternoon session, lunch has been eaten and everyone is tired. You are covering building materials and the going is rough, people nodding off right, left and centre. Let's play the ball game!

Get the group into a circle and hold an imaginary ball in your hand.

If you throw it straight across the circle, say 'Wheesh', if you pass it to the person on your right or your left, say 'Whoosh'. Tell them that if they don't know where the ball is, they should throw both hands up in the air, wave them about and say 'biri, biri, biri, biri...'

It'll raise lots of laughs, people will liven up, and forget that they're tired. But you will need to change the tempo soon, or give a good long break.

1

Earthaven Ecovillage using recycled materials and powered by the sun.

Chapter 2. Permaculture Tools

Several of the founders of the Global Ecovillage Network were Permaculture teachers and designers, and this has given GEN a strong connection to Permaculture design. We could argue that Permaculture is the science and theory, while Ecovillages are the art and application. Whichever way we might want to look at the connection, there is no doubt that Permaculture is a valuable set of design tools for planning Ecovillages.

Permaculture is about designing sustainable human settlements. It is a philosophical and practical approach to land-use – integrating microclimate, functional plants, animals, soils, water management and human needs into intricately connected, highly productive systems. Permaculture looks for the patterns embedded in our natural world as inspirations for designing solutions to the many challenges we are presented with today.

Permaculture was developed in Tasmania, Southern Australia in the late 1960s and early 1970s by a collaboration between a professor of ecology, Bill Mollison, and his student, David Holmgren. The idea arose that modern agriculture, indeed nearly all plough agriculture, was creating soil loss and depletion, and that we needed to develop an agriculture that was permanent rather than a temporary soil robbery. Thus 'Perma-nent agri-culture' became 'Permaculture'. This was essentially a response to pollution and bad agricultural practices, and the solution was found to be in the radical idea that the patterns found in natural ecologies contained all the patterns we needed for redesigning our own systems.

Mollison travelled the world, spreading the name, which has enough of the snappy quality about it to be appealing to many cultures and works in many languages. During these travels he developed the standard foundation course, known as the Permaculture Design Course (PDC), a 72-hour course upon which this book is based. In his travels, he left behind a series of projects, educational centres, and later Ecovillages that during the late 1980s and early 1990s started communicating with the cheap and simple tool that was then developing: the internet and email.

The international Permaculture network was born, PDCs proliferated in many countries and in many languages, and solutions to the problems that had confronted Mollison and Holmgren a couple of decades earlier started to proliferate.

Permaculture definitions

As a Permaculture instructor, I often get asked to explain what Permaculture is, so it's useful to have a concise explanation handy:

Permaculture is a set of design tools, based on observing the patterns in nature's cycles, which we can use as models to design the infrastructure we need for a sustainable future.

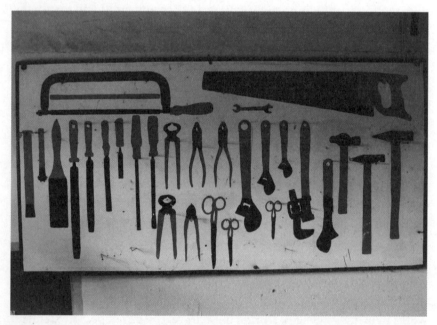

Permaculture is a set of tools, it helps to have them clearly available.

Let's take it apart and explain each element.

- Design tools = ways of thinking, thought patterns.
- Patterns in nature = this is the most innovative side of Permaculture, and indicates that the patterns we perceive in nature are real guides for our own designs.
- Infrastructure = everything that we need; food, shelter, transport, economics and culture.
- Sustainable future = caring for future generations.

2

Let's introduce some definitions that have a bit more zest. They can best be understood by those who are already familiar with the basic definition. They are great starting points for discussion in small groups:

'Permaculture gives no answers, it is a series of questions leading us into the future.'
Because every person and every situation is different and unique, we cannot give set answers, but we can empower people to find the solutions that fit their situation.

'Permaculture is eco literacy.'
We have largely lost the ability to read the book of nature. Permaculture encourages us to interpret the signals that nature presents in front of us, and use these interpretations to act wisely.

'Permaculture is a dialogue with nature.'
Whenever we do something in or to nature, we should observe carefully. What effect does this action have? What changes are happening as a result of it? Our observations should help us modify our next action, and in this way we can achieve a dialogue, a two-way conversation, where each action is followed by a response.

'Permaculture is action focussed.'
Even though founded as a body of theory, Permaculture includes actions, even if these are designs or plans to be implemented in the future. When teaching building or gardening, no course would be complete without

some hours spent out in the field or on the building site, working hands-on with the materials. Indeed, it is the dialogue between our hands and the material we are shaping that is the content of the dialogue just mentioned.

'Permaculture takes responsibility for the future.'

Everyone can do something, and it is the finding of the 'something' that you can do which is the aim of the course for you. This is just another way of saying that 'Permaculturalists don't attempt to do what they can't do' or 'Permaculturalists make do with what they have'. We know that if every person does one little thing that they can do successfully, the whole world may change. Recognising that you have responsibility for your actions, and finding your task in the world, is one of the most meaningful things about Permaculture.

'Permaculture is pragmatic rather than dogmatic.'

There may be lots of rules and principles, but in the end, each one of us finds his or her self in a unique situation, and it's impossible to lay down a law for each one. Of course we would prefer to practise natural gardening with soil, compost and rainfall, but there are many people in the world who find themselves in places where these three are not to be had. So then we suggest hydroponics, acquaponics or even aeroponics, anything that will solve their problem of food access.

'Permaculture ethics are – Planet care, People care, Sharing resources.'

Now Mollison was never the big one on philosophy and ethics, but these three simple ethical thoughts are good enough to be shared by most social philosophies and all the major religions in the world. In advanced courses we may delve into the ecosophy of Arne Naess, the scientific worldview of Johann Wolfgang von Goethe, the social anthroposophy of Rudolf Steiner, or the 'thinking machines' developed by Patrick Geddes. But for the foundation course it is really enough to suggest that there is an ethic inherent in Permaculture, and ask participants to take it seriously.

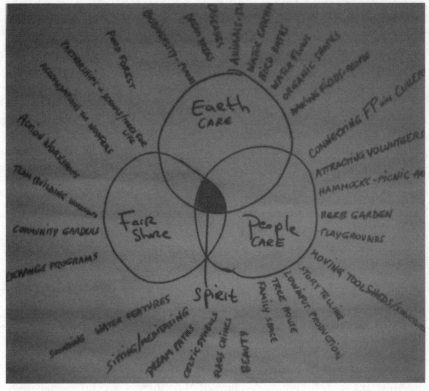

2

Where the three Permaculture ethics overlap.

Biomimicry

Mollison didn't have a very positive relationship with the academic world, and Permaculture has not really been accepted as a subject for academia. The idea of using nature as a model has, however, and so we find permacultural ideas under the title *Biomimicry*. This seems to sit well with engineering and product design.

Biomimicry believes there is no better design partner than nature. But biomimicry is more than just looking at the shape of a flower or dragonfly and becoming inspired. Biomimicry is now an accepted subject for academic research and teaching, and shares with Permaculture that central concept of nature being the best teacher. Of course, this idea would not have been viable until we began to accept that ecology is a subject in its own right. This happened during the 1960s. Rachel Carson's book *Silent Spring* was highly influential in bringing the subject into the limelight, but Eugene Odum was perhaps the academic who actually laid the foundations of ecology as a subject. Mollison is very aware of this and acknowledges a large debt to Odum.

Take a look at any natural ecosystem, such as a prairie, and you will see a remarkable system of food production: productive, resilient, self-enriching, and ultimately sustainable. The modern agricultural practices of humankind are also enormously productive, but only in the short term: the irrigation, fertiliser, and pesticide inputs upon which modern food crops depend both deplete and pollute increasingly rare water and soil resources. Biomimicry has been working successfully to revolutionise the conceptual foundations of modern agriculture by using natural prairies as a model: they have been demonstrating that by using deep-rooted perennials in agricultural systems which mimic stable natural ecosystems we can produce equivalent yields of grain and maintain and even improve the water and soil resources upon which all future agriculture depends.

Looking at another example, we generally think of termites as destroying buildings, not helping design them. But the Eastgate Building, an office complex in Harare, Zimbabwe, has an air conditioning system modelled on the self-cooling mounds of *Macrotermes michaelseni*, termites that maintain the temperature inside their nest to within one degree, day and night, while the temperatures outside swing from 42°C to 3°C.

Permaculture principles

David Holmgren developed twelve principles that he elaborated upon in his book: *Permaculture: Principles and Pathways beyond Sustainability*, published in 2002. These have subsequently formed the principles that many Permaculture instructors now teach.

- Observe and interact
- Catch and store energy
- Obtain a yield
- Apply self-regulation and accept feedback
- Use and value renewable resources and services

- Produce no waste
- Design from patterns to details
- Integrate rather than segregate
- Use small and slow solutions
- Use and value diversity
- Use edges and value the marginal
- Creatively use and respond to change

2

Observation is a fundamental Permaculture principle.

Design as a divine art

The design process has some really magical qualities. It bridges over two dimensions, from the non-material or spiritual, to the material.

Whenever I have an idea or a thought, this does not exist in the material, physical world. It is composed of thoughts, feelings, associations and intuitions that arise within myself. Where they come from is a matter of opinion, some believe they are the result of genes, instincts and influences from the environment, others that they arise from the 'spirit'. The jury is still out on this one. Wherever these ideas come from, I can share them with others, and if enthusiasm is aroused and we start working together, a group thought will arise.

A couple of years ago we had the idea of building a compost toilet on our property. We raised the suggestion with the first group of volunteers who came to stay with us that year, and started planning and designing. One of the first things we had to do was to research composting processes to do with humanure, something that then entailed changing our composting process. We began clearing the room that was to be used for the toilet, and set ourselves the challenge that we would buy nothing, using only materials we had to hand. As the summer progressed, different groups came and went, and the project ran like a relay race, each group taking over where the last group left off.

A compost toilet designed and built by a succession of volunteers over the course of a summer.

At the end of the summer the toilet was in use, the compost heaps had been rearranged, and the toilet itself painted, with instructions and information in Norwegian and English, with a bookshelf of good books and even a reading light. An idea, which had been slumbering in my mind for a couple of years, was manifested in the material world by a group of people working together, many of whom did not even meet each other!

For those who read the Bible, in the first book, in Genesis, God is creating the world. 'Let there be light!', 'Let there be land!', and so on. The story implies that God is creating something out of nothing. Whether we choose to believe or not, the image is a powerful one, God as the Creator. My feeling is that when we take in use the Permaculture design process, we are as gods, we are following the route of an idea, which originates in some other dimension of thought or feeling, and we are instrumental in manifesting this in the physical world.

'Let there be a compost toilet! And they saw that it was good!'

By concentrating on the idea of design we can find a new way of thinking about

Permaculture means working together with others and with the earth.

the world, and of coming up with imaginative solutions to the challenges that we are confronted with. Design is the manifestation of a new holistic, scientific model corresponding to a new worldview of interdependence. Permaculture design is a way of integrating different dimensions of life, different sciences and traditions, both the inner invisible and the outer physical.

Permaculture design is action oriented, where citizens, participants and planners can help value-based change happen by sharing a worldview of oneness, a vision that allows evolution to continue, and creating a world that works for all. Permaculture design embraces all levels of society.

Permaculture in the world

The Permaculture Design Course (PDC) is the structural backbone of the Permaculture movement. Anyone who wants to be part of that movement will sooner or later take the PDC and get the certificate. The course lasts a minimum of 72 hours, and has a pretty standard programme that includes most fields of human activity: food, shelter, energy, water, economics, culture and society. Just look through the chapter headings in this book and you'll get the idea. To get the certificate, participants have to present an individual design project.

Individual design work is expected from each participant on every Permaculture Design Course.

For those who have the certificate, and want to aim higher, there is the possibility of gaining a Diploma, which is awarded by local or regional

Permaculture Institutes. The rules are slightly different according to which country you live in, but basically it is given after at least two, often many more, years of active work within a Permaculture field. This could be community or project development, education, or any of a number of other themes.

International Permaculture Convergence in Europe in 2005. Mollison and Holmgren are both in the middle at the back, wearing blue shirts. The author is furthest on the right.

There are regular meetings at local levels, national associations, regional meetings, and continental (European, for example) and the occasional international gatherings. These are often called convergences to emphasise that they are not hierarchical decision-making meetings, but a gathering and sharing.

Transition towns

Over the years, Permaculture has created new initiatives which have gone on to develop by themselves, to grow, to change, and even break out of the purely Permaculture network. The largest and most well established is the Global Ecovillage Network, created by a group of Permaculture designers, mentioned in the last chapter. The most recent is Transition Towns (TT).

TT was started by Rob Hopkins, a Permaculture teacher living in Ireland in 2004 and increasingly concerned about the issue of peak oil. Of course it was obvious to many people that we were using oil much faster than nature was creating it, and that sooner or later we would find ourselves with very little oil left. In the early years of this century, researchers began looking in detail at this process and found that oil extraction follows certain very clear and repeatable patterns, resembling the so-called bell curve, and that as we go over the peak of the bell curve, so our extraction of oil will inevitably decline. Their research showed that we were very close to the peak exactly at this time. We have been extracting oil for about a century, and will have another century of use until only insignificant amounts can be extracted. This spurred a number of Permaculture designers and groups to looking at how we can cope with a future using less oil than we do now, bearing in mind that we are totally dependent upon cheap oil for most of the things we need.

Hopkins moved to Totnes in Devon, began showing films and talking to people about a future without oil. In 2006 he launched an event that he called The Unleashing, to find out what ordinary people can do in the face of major changes. Out of this came the idea that local communities can design a transition to a future with less oil, and that if they do so consciously and collaboratively, this will be far less dramatic and painful than having such a future thrust upon them by outside forces. The Transition Town movement was borne, and Totnes became the first Transition Town.

In many ways, TT is very like the Ecovillage movement, designing a sustainable future that is based on small local communities. The big difference is that instead of designing a new community from scratch, TT begins with an existing community, working through existing social organisations; schools, kindergartens, local associations, and of course the municipal authorities.

Central to TT is the idea of resilience. This term comes to us from ecological science and is about the ability of a system to hold its shape in the face of change and outside pressure. Resilience describes the ability of a system to maintain itself, in this case maintaining a social cohesion where not just survival, but growth and development will be the result of positive cooperation and mutual caring between the residents and members of the community.

TT has its own training programmes, and there are sets of guidelines and official membership criteria. An interesting detail is that every officially recognised TT group should have at least one member who has taken the Permaculture Design Course.

There are now about fifteen hundred official Transition groups in over forty countries, but there are many more unofficial groups operating.

The changing face of Permaculture

So what is Permaculture? It's a design tool, a body of knowledge, an international network, and a process of personal transformation. It's composed of many tens of thousands of people, and their synergy combines to make it much more than just a collection of individuals. It is an organic living entity of its own, a part of what Paul Hawken so admirably describes as the planet's immune system in his excellent book *Blessed Unrest*. As a living organism, it displays the characteristics of other creatures; it grows and changes, it develops and matures. Like plants, it flowers, sets fruit and scatters its seeds.

I spent the first half of 2008 in Israel, teaching Permaculture and watching the Israeli Permaculture network change from a pioneering effort to an established mature organisation. Just as puberty in human beings is a difficult age, so the process I was watching was not always easy.

During the previous decade hundreds of people in Israel had taken their Permaculture Design certificates. There was a network of highly varied projects.

There were some sharply divided differences of opinion, between those who felt they needed a defined structure, and those who had no time or patience for discussing rules. Some of the questions that arose were: Who can use the name Permaculture? To whom does it belong? Who is qualified to teach? Who can give Diplomas and what criteria are there? These are serious questions, and need to be asked wherever Permaculture is being used, taught and developed. These questions bring us back to a definition of Permaculture, and how we see it developing and growing.

Israel is in a precarious situation where war, environmental problems, internal social conflicts and problems of water as the rainfall becomes more erratic and less reliable, all contribute to a sense of impending disaster. The question soon came up: How can we waste time discussing rules when the country and the region is fast becoming a disaster area? Is it ethical to get involved in bureaucracy when people need immediate practical help to solve wastewater problems, energy misuse and indeed lack of food? These questions should not be dismissed lightly if we are to keep a cohesive Permaculture movement. The last thing we need is conflicts that take up time and energy, *really* wasting our resources!

Luckily Permaculture is based on using natural ecological templates. Whenever we need an ethic or come up against problems and conflicts, we can go back to our basic principles and take our inspiration from nature. This is not so much a philosophical question of right or wrong, good or evil. One of the basic Permaculture tools is looking for patterns in nature that can help us design our way out of problem situations.

When we look at succession in nature we find that even established ecologies still contain pioneer plants. When a tree falls in a mature forest, and the canopy is opened up to allow sunlight to penetrate, these pioneer plants lying dormant in seed and root sprout forth, soaking up the sunlight and creating new biomass and a wide variety of niches. Nature needs diversity to deal with a wide range of situations. The principle of succession is not that each successive ecology totally replaces the preceding ones. Successive guilds (that is, groups of compatible plants) add to existing ones, they dominate but without totally removing previous plants and guilds. Pioneer plants are always present. It is not an 'either – or' situation, but an 'also – with'.

What can we learn from this?

As the Permaculture movement worldwide grows and establishes mature social ecologies, we find that we still need a wide variety of activities. We must not fall into the trap of completely replacing the pioneering plants with the trees of the old growth forest. Structured quality control needs to be combined with pioneering hands-on practical projects. Mulch gardens and domestic grey water treatment design needs to go hand in hand with Permaculture Diploma holders enforcing the criteria to get a PDC certificate. We need to keep our sense of youthfulness and frivolity, but at the same time be taken seriously by the bureaucrats of mainstream society.

This can be seen as a question of language. As Permaculture is seen to be successful, mainstream planners begin asking questions. In many places around the world rural and urban planners are turning to Permaculture designers and Ecovillage planners for help in creating more mainstream zoning and building regulations. We need to learn their language in order for our Permaculture strategies to be adopted. Quality control is part of that language. We need to be able to convince them that someone who has a PDC certificate has gained a working knowledge of Permaculture, and that a Permaculture Diploma is proof of serious technical skill and experience. Occasionally it's tactical to put on a suit, bring in the PowerPoint presentation, and convince mainstream planners that we have relevant answers. It's not enough to concentrate only on the hands-on projects, disregarding the documentation of successful projects, serious practical research, building codes and zoning regulations.

As always in Permaculture, when in doubt, when a conflict arises, when problems seemingly overwhelm us, the most valuable tool we have is nature herself. The ethics of quality control lie in the basic Permaculture strategy of looking to natural systems for inspirations and solutions and using the patterns we see there in order to design our own systems.

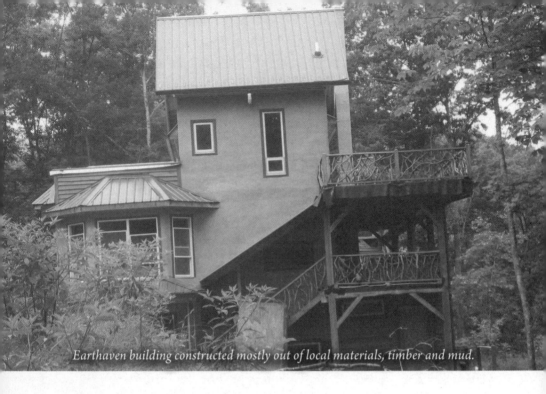

Earthaven building constructed mostly out of local materials, timber and mud.

Community profile

Earthaven

Earthaven Ecovillage was founded in 1995 by a group of Permaculture designers and is located in North Carolina, USA.

It is spread out in the forests along a valley in the steep Appalachian hills, with several neighbourhood clusters, a co-housing building with several families, a seminar centre offering courses in Permaculture, and several small businesses.

The forest is natural re-growth after clearing in the early years of the twentieth century, and is again being cleared for construction, and replaced by buildings, gardens or fruit and nut trees. Re-afforestation is being carried out on a big scale, and according to Albert Bates, Earthaven is more than carbon neutral, and an example that it can be done.

Many techniques of ecological building can be found there, strawbale, mud, timber and junk. They are well insulated, and heated by solar, backed up by wood-burners using material from the forest.

2

Inside the community hall.

Hillside stabilised with tyres, and planted with vegetables.

Electricity is from photo-voltaics, water is from wells, springs and rain catchment. Grey water is recycled, and the toilets are composted.

Earthaven is engaged socially in several of the social movements such as Occupy, and hosts and teaches seminars on a wide range of subjects, such as Permaculture, community, Integral Theory, Ecovillage design, and building, gardening, planning and social aspects. In many ways Earthaven can be seen as a showcase for Permaculture techniques.

For more information: http://www.earthaven.org

Tips for facilitators and groups

INTRODUCING Permaculture principles

I use the twelve principles developed by David Holmgren (see also above).

- Observe and interact
- Catch and store energy
- Obtain a yield
- Apply self-regulation and accept feedback
- Use and value renewable resources and services
- Produce no waste
- Design from patterns to details
- Integrate rather than segregate
- Use small and slow solutions
- Use and value diversity
- Use edges and value the marginal
- Creatively use and respond to change

Instead of teaching these principles as they stand, I introduce the group to a random design exercise. I have two envelopes. One contains elements, at least twelve different ones, the other at least six prepositions (linking words). I divide the group into not more than six smaller groups. Each group selects two elements and one preposition. This could be anything from 'Tree beside House', pretty straightforward, to 'Road under Herb

Garden', more of a challenge. The participants now have twenty minutes to sketch out a design. Each group then has five minutes to present their design.

As they present, I look for examples of the various principles that Holmgren developed. At the end of all the presentations I then introduce the twelve principles to the participants, referring to specific designs by specific groups. In this way they experience the principles as a recognition of ideas they had already been using, and feel more of an ownership.

2

Plenty of space around
the houses at Dyssekilde
Ecovillage in Denmark.

Chapter 3. The Size of Communities

Before we can use our Permaculture design tools we need to ask some basic questions about the Ecovillage project we are focussing on. I am assuming that you are part of a group who want to design and build an Ecovillage. Three of the most fundamental questions you need to ask are:

- What size are you envisaging?
- How much privacy, and how much integration do you envisage?
- Are you building on a greenfield site, or are you retrofitting existing buildings, perhaps with residents already living there?

Because the size and social character of the community will demand specific design approaches, these questions must be addressed first. So we will take them in order.

Size of Ecovillages

In the original definition of Ecovillages, suggested at the GEN conference at Findhorn in 1995, Diane and Robert Gilman put forward the following when addressing the issue of size: '... somewhere between fifty and five hundred members, but with exceptions.'

In the nearly twenty years that have passed, the limits suggested by the Gilmans have been stretched.

Auroville in India is often reckoned to be the largest, a check on their home page in January 2014 finds a population of 2,160 people, and growing. In many ways, Auroville can be seen as a collection of smaller communities or neighbourhoods, all gathered under one umbrella. Perhaps we may one day start talking about Ecotowns or Ecocities. Certainly Auroville is one of the largest, and has ambitions to grow a great deal more. They are aiming at well over 20,000 residents.

The other end of the scale I found in a film released by L.O.V.E. Productions in 2010, 'A New We'. This film presented short clips from ten Ecovillages in Europe, one of which, Finca Tierra in the Canary Islands, totalled one full-time resident, and a group of temporary volunteers. It would be difficult to find a smaller Ecovillage.

As you will realise, there are wide divergences here, and it's up to you as a group to choose.

Aerial photo and map of Dyssekilde Ecovillage in Denmark.

Privacy and Integration

I lived for sixteen years on Kibbutz Gezer in Israel, where for most of that time we ate our meals in a central dining hall, had our laundry done in a central laundry and our children spent most of the working days in the children's houses. This meant that our houses could be quite modest, no need for washing machines or large kitchens and dining areas. Our houses were tiny, and in the terminology developed by the one hundred years of Kibbutz development, many people still use the word 'room' for their house, harking back to the time when that was what there was, a room for each individual or couple.

Many Ecovillages today are modelled on the co-housing pattern, where each family is responsible for their own finances, food, washing and childcare. There is usually a central house where the community can come together for meetings, meals and festivals, maybe a childcare facility, and of course if someone wants to provide laundry services, there is nothing to stop them. However, each house will have to be planned with all the services that a family needs. At Forgebank Co-Housing near Lancaster in the UK they have saved a considerable amount of private space by incorporating a children's play area, essentially just a couple of rooms full of toys, a communal laundry and bike shed doubling as a workshop.

It will be obvious that the amount of integration will determine many of the planning features. It's quite useful to be clear about this in order to avoid costly retrofitting later on. Sometimes this is unavoidable, if the ideology of the group changes.

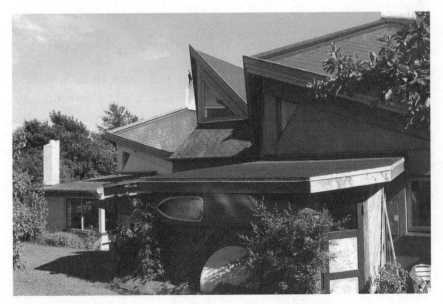

Each family is responsible for building and maintaining its own home in Dyssekilde Ecovillage in Denmark.

The kibbutz movement began before World War One, and as it grew through the 1920s and 1930s, it became the norm that children slept in children's houses. Indeed, this was a defining feature of the movement and a great deal was written about it, including in the 1960s, Bruno Bettelheim's *Children of the Dream*, a landmark book about child psychology and sociology. When we joined our kibbutz in 1984, the children's houses had closed as a night time option a year or two before, and children were sleeping at home with their parents. This practice spread throughout the 270 kibbutz villages during the 1980s, entailing a massive building programme as houses had to be expanded to include extra bedrooms. Many see this building programme as one of the factors that escalated the total kibbutz debt during the 1980s.

I know of very few communities that have children sleeping in children's houses today, and this story from the kibbutz may be an extreme example from the past, but the principle remains just as important, that the amount of integration will determine many of the planning aspects, and have economic consequences. There is no blueprint, you as a group will have to choose. Better to do this early on, so that your planning can reflect the kind of community you want, and you can budget accordingly.

Greenfield or retrofitting

With the planning and zoning regulations that seem to be putting a stranglehold on initiatives throughout the western world, a greenfield site is less likely to be a starting point for a new Ecovillage. Of course, this may change as zoning regulations are updated. Right now, it seems that zoning regulations and planning laws reflect an old, outworn paradigm of compartmentalising, rather than the integrated approach used by Permaculturalists and others who are looking to nature as their model. One idea that has emerged is to request that the planning authorities release specific small areas from all planning regulation, in order for Ecovillage projects to experiment with new ideas currently rendered illegal.

Until that happens, taking on an existing group of buildings may be a lot easier. Sometimes things really fall into our laps. Old abandoned army bases, schools, hospitals and kindergartens may be a gift to those of us who can find them and get hold of them. Both ZEGG and Lebensgarten Ecovillages in Germany are two such examples. ZEGG was established in 1991 on a disbanded Stasi training camp, while Lebensgarten was originally built by the Nazis as a slave labour ammunition factory, then used as a British Army camp, and lying derelict for some years before the Lebensgarten group bought it in 1985.

3

At ZEGG an old Stasi training camp has been transformed, here the organic permaculture gardens.

The Ecovillage at Findhorn is built on an old caravan park, and they later acquired a large hotel in Forres. Many of the hand-built houses are officially temporary mobile homes in order to get round building regulations.

Ecovillage initiatives in Senegal are based on upgrading traditional villages straight into a post-industrial eco sustainable culture.

Here in Norway, where there is no strong tradition of villages in the European sense, many Ecovillage type projects are centred on an existing farm, and entail adding more houses to accommodate a modest group of people. Again, building and zoning regulations create a very difficult situation for groups to operate in, for instance, the Ecovillage at Hurdal spent a decade negotiating their way through a jungle of red tape.

There is no set recipe, every situation is unique.

Community as organism

If we are to read our culture from the mainstream media, it must be obvious that it is a very *yang* type culture, dominated by masculine qualities such as aggressiveness, greed and domination. Large global structures such as the World Trade Organisation, the World Bank, transnational corporations, and growth economics control our lives and squeeze out the local civil society.

Probably many of the readers of this book also have at least a foot, many much more, in the so-called alternative, new age or new paradigm culture. This is a much more *yin* culture focussing on how we live with each other within nature. This entails, amongst other things, having an open decision-making structure where each person is listened to, and creating social frameworks that give opportunities for personal initiatives.

On a planet that is as interconnected as ours, we need a balance between the *yang* of globalisation and control, and the *yin* of locality and personality. We need a worldview of Oneness and global interdependence.

Every group of human beings that is gathered around a focal point forms a fellowship, be it stamp collecting, football or ecology. Living together creates a stronger glue, the loose gathering becomes firmer, the character of the fellowship easier to comprehend. Every community has its own biography, its own history, traditions and often language, and a shared set of cultural characteristics. In this way a community can be seen as an organism, obeying laws and rules, ways of development, and even with its own spiritual and cultural traditions. Every Ecovillage is a unique individual.

In some communities this is quite explicit, and the Camphill network has even developed a group within it called The Camphill Community, members who meet regularly to take responsibility for the spiritual development of the fellowship. They have no legal definition, no real budget to call their own, no mandate from any other group within

Camphill, but meet regularly, both locally, regionally and internationally, and apply themselves to nurturing the spiritual life. Despite not having a firm legal existence, suggestions originate in the Camphill Community, and Camphill is well known for its chapels and halls, expressions of both community endeavour and spiritual values. In 1990 Joan de Ris Allen published the book *Living Buildings,* a celebration of fifty public buildings from various Camphill communities throughout the world.

3

The Kristoffer Hall at Camphill Vidaråsen, designed by architect Joan de Ris Allen.

Many Ecovillages are built around core spiritual values that have their expression in sacred buildings, chapels, meditation places and shrines. Perhaps the most striking is the underground temple built by the members of Damanhur. Located in the Alps, not far from Turin, Damanhur was founded in 1977, and today comprises over six hundred people spread through several communities. They have numerous workshops, their own currency, representation on the local municipal council and even their own fire brigade. Damanhur is best known for its underground 'Temple of Humanity', a vast 6,000 cubic metre network of tunnels and caves that they have excavated themselves.

Other Ecovillages are more secular, and their expression of community in the architectural sphere may be a meeting hall, a common house

Ritual outdoor areas at Damanhur.

or the shared spaces that are necessary for any community to be able to have all their members in the same place.

All these aspects have to be taken into account when planning the Ecovillage. Every type of society will express itself in the material buildings and infrastructure that it creates.

Industrial society manifested itself through such phenomena as suburbs, factories, parking lots and high-rise business-district office blocks. Now that we realise and understand the damaging impact of industrialism on the planet it's becoming clear that a more sustainable culture will have to replace this. This is the essential message of this whole book, and probably why you are reading it. The new emerging sustainable culture will have to manifest in quite other ways, building other structures, redesigning our habitat. Some of these iconic structures might be windmills, living roofs (roof gardens), cycle lanes taking up more space than roads, and open spaces filled with forest gardens producing food.

Part of your challenge as a design group will consist of defining these structures, planning them, and building them. What would they consist of? How would they look? How would you build them? What materials would you use? In order to keep on track, you might like to ask yourselves the question: How does this design manifest and reflect our core ecological, ethical and social values.

Community profile

Auroville

A profile by Ann Ellsie Morger

Auroville was founded in 1968, inspired by Sri Aurobindo (1872–1950) from India and The Mother, born Mirra Alfassa (1878–1973) from France. Sri Aurobindo had studied in England to be a lawyer. He had considered joining Mahatma Gandhi's independence movement, but felt that he had to take another path in life. The Mother had a dream that took her to India and to Sri Aurobindo. They were both mystics with a background in Yoga and Meditation.

Auroville lies about 160 km south of Madras in Tamil Nadu in South India, not far from Pondicherry. There are about two thousand inhabitants from around forty different nations. 40% of them are Indian, 20% are French, and 20% German. The last 20% are from other countries. There are about one hundred settlements in an area of about twenty square kilometres. The whole project is designed for up to 20,000 to 60,000 residents. The area was originally desert, but has been forested and is now green and fertile.

In the middle of Auroville stands the Matrimandir, with an enormous gold dome. This is the spiritual centre of the Ecovillage. The Matrimandir was started in 1971 and finished in 2003. The aim is that it should be a spiritual centre for everyone in Auroville. The Inner Chamber is the heart of Matrimandir and here people can sit in meditation or concentration. To experience stillness, with many people from different countries and cultures sitting around the crystal ball, is highly magical. The aim of Auroville is to help each one raise his or her consciousness, for the good of all humanity, based on the following principles:

- To live in Auroville, one must be a willing servant of the Divine Consciousness:
- Auroville is a place of constant education, of constant progress, and a youth that never ages.
- Auroville wants to be the bridge between the past and the future.
- Auroville will be a site of material and spiritual research for a living embodiment of an actual Human Unity.

Auroville is developing an internal economy that will eliminate the internal use of money, and that ensures every person's needs are met. Everyone receives food and accommodation, and some pocket money every month. If they want to travel or attend courses outside, each one has to earn the means themselves. The aim is to be self-sufficient in food, clothes, schooling and health services.

They grow a good deal of their own food organically. They have their own school, several businesses in the fields of organic food, clothing, leather goods, computer services and book publishing. The health centre is used by residents in the area, and from neighbouring towns. They have their own administration, with a Town Hall and a Visitors Centre.

Auroville is recognised by UNESCO, and receives support from all the Indian states, and several countries throughout the world.

Read more on: http://www.auroville.org

Catching the solar energy. Photo: Laura Podoski

Tips for facilitators and groups

Whole community

This is a good exercise to introduce the concept of privacy and integration. You will need a board or flipchart.

Start by writing 'Whole Community' inside a circle at the centre. Divide this into three broad areas: Personal, Social/political and Ecological.

Now go round the circle adding in the words: Ethics, Self-development, Interpersonal, Political, Group dynamics, Education, Building, Farming, Business, Infrastructure, Wilderness experience, Religion. You can ask the group to contribute words rather than just writing from this list. Use the mind-mapping technique of creating more nodes as you go along.

You might then initiate a small group discussion session, for around fifteen to thirty minutes, to assess the issues of size and privacy/integration on specific issues outlined on the board. In the plenum following, you might ask for a general discussion rather than a report from each group.

A great deal of food can be grown in a small space working by hand.

Chapter 4. Farming

We won't get very far without food, so let's start right here. How can we grow and distribute food in fellowship? Farming has as one of its main aims to produce food for people, and must be integrated into social and economic systems. In this chapter we will look at some of these systems and how we can improve them.

Small-scale farmers

According to the United Nations Environment Programme's report 'Towards a Green Economy', published in 2011, small-scale farmers produce food for 70% of the world's population. This was subsequently confirmed by a later report from United Nations Global Impact that stated that small-scale farmers produced 70% of all the world's food.

There is a myth that has been widely disseminated that industrial-chemical farming is the most efficient way of producing food, but this is not borne out by facts. It is true that when we measure productivity against labour, the worker in the big tractor does produce more food per hour than the man or woman with the hoe. Whether this takes into account the whole infrastructure, the factories producing tractors and parts, the oil industry delivering diesel, the trucks, ships and supermarkets moving the produce from field to homes, remains to be seen.

However, we live in a world where there are plenty of people, and only a finite amount of farmable land. When we measure produce per acre, it turns out that the smaller the farm, the more efficient it is. According to the United Nations Department of Economic and Social Affairs in 2011, small farms produce between 20–60% more food per acre than large industrial-chemical farms. The United Nations Food and Agriculture

Large tractors need a lot of space to manoeuvre.

Association found in 2007 that organic farms worldwide were 132% higher yielding than the world average for all agriculture!

The conclusion is that the future of world farming, and the key to feeding the world's population, lies in the small family farm, using organic methods. The way forward is by education through doing, encouraging as many people as possible to make compost, build soil and grow vegetables. From herbs and salads grown on your balcony or outside your kitchen door, to family vegetable plots and allotments, to community gardens and Community Supported Agriculture schemes, to family and community farms cultivated organically by people who care about soil and plants, these are the keys to creating food for every person on the planet!

Individually, small farmers are weak, especially when confronted by global corporations. But there are so many small farmers in the world that collectively they can be a force for change. La Via Campesina (The Peasant's Way) was founded in 1993 in Belgium and now numbers about 200 million small-scale growers, fisherfolk, farm and land workers around

Harvest festival at Camphill Solborg in Norway, displaying vegetables grown during the summer.

the world. They have demonstrated in their thousands against the World Trade Organisation, the World Bank and the United Nations when these have met to discuss food and agriculture.

One of the main issues for La Via Campesina is food sovereignty, which they defined in a declaration at the Forum for Food Sovereignty held in Selingue in Mali in 2007:

> Food sovereignty is the right of peoples to healthy and culturally appropriate food produced through ecologically sound and sustainable methods, and their right to define their own food and agriculture systems. It puts those who produce and consume food at the heart of food systems and policies rather than the demands of markets and corporations. (Reported in *The Guardian Weekly*, June 2013.)

One key element in establishing food sovereignty is to look after the seeds. Smallholder farmers have been into precision agriculture for millennia. In India fifty years ago there were an estimated 100,000 landraces, or local seed varieties. Today fewer than 6,000 are grown every year. The traditional varieties perform better in marginal conditions, and are based on careful observations by generations of farmers. These varieties can be both drought- and salt-tolerant, important factors in the changing climate of the present.

One of the problems for small farmers in India is that farming is not really regarded as a business. A few years ago 'I Say Organic' was established by Ashmeet Kapoor in order to create direct links between organic producers and consumers. He states that 'properly practised organic farming reduces input costs for fertilisers, pesticides and seeds, dramatically improves farmer health and enhances the fertility and resilience of their land.' The organic sector is growing fast in India, with a number of Indian states, including Mizoram, Uttarakhand and Sikkim considering going 100% organic. (Reported in *The Guardian Weekly,* March 2012.)

Forest farming

A food forest mimics a woodland ecosystem by combining a number of different plants at different levels. Fruit and nut bearing trees form the top level, under which we establish berry shrubs, edible perennials and annuals. Companion plants are included for natural pest management and legumes provide nitrogen and mulch.

Trees provide valuable functions in holding the whole system together by holding water, cleaning the air, stabilising soil and rocky slopes, creating wind breaks and shade. A forest is a single assembly, a single organism, not just a collection of multiple units. A young forest needs high input and gives low yields. A mature or old forest needs only low inputs and gives high yields. A forest with clearings will contain all levels, and creates multiple edges.

We can design a mixture of trees and other plants giving abundant yields by using the concept of stacking. At the top we have large canopy trees, these might be concentrated towards the north in the northern hemisphere, creating sheltered spaces below them for smaller fruit and nut yielding trees and berry shrubs below them again. Herbaceous perennials and ground cover plants such as strawberries will fill the spaces between the larger plants, and there are levels below ground also to be considered, where we might put root vegetables, or encourage mycelium that may give us edible mushrooms.

4

Alvastien Permaculture Farm on the West Coast of Norway, integrating trees, buildings and gardens on very steep land.

When designing a forest garden concentrate on fruit, nut and leguminous (nitrogen fixing) trees. Avoid large timber trees, as felling can damage nearby plants. Pruning should give plenty of small firewood. Companion planting and diversity is important for pest management. Checking with local conditions and traditions is a key to success. In our modern arrogant thinking, we are sometimes led to believe that old traditions were rubbish, and that we know so much more today. There is a great deal of wisdom in traditional thinking, and Permaculture has as one of its principles that we 'should honour the elders'. How did people grow things here in the past? What kind of varieties did they cultivate? What kind of techniques did they use to solve the challenges that this particular environment threw at them? By asking ourselves pertinent questions such as these, we may end up learning a great deal. Permaculture is about asking the right questions that will lead us to good answers.

Community-Supported Agriculture

Today there are very few people involved in food production in the west, but we have no choice whether to farm or not if we want to eat. We either have to support farms where other people do the work for us, or we have to do the farming ourselves. In the economy that is practised in the west today it is difficult for farmers to make a living without massive government subsidies. Will the idea of cooperation and mutual support give a new inspiration to agriculture? There is a desperate need for a new set of economic relationships, a new social community development.

The idea of a new relationship between the grower and the consumer arose in the 1980s as a reaction to the growing food miles, and the increasing role of the supermarket forcing its way in between the farm and the kitchen. By 1990 there were estimated to be about sixty Community Supported Agriculture (CSA) schemes operating in the United States. By 1997 this figure had topped the 1,000, and it was estimated that over 100,000 households were involved. Since then many more have been established, and it seems to be a growing trend in many countries. Here in Norway there were about nine schemes operating in 2013, by the summer of 2014 there were fifteen, with long waiting lists and more CSA schemes being planned. The established food industry has not yet seen this as a threat, but as more organic food is sold in the supermarkets, the price will come down, and it may well be that CSA type schemes will find it hard to compete financially.

A CSA can be defined as a system of farming where the producer and the consumer share the risks and the benefits. There is a whole spectrum of approaches to choose between, from the least integrated model of the classic farmer's market to the most radical, the community-owned farm.

Farmers' markets have the least involvement or integration between the grower and the consumer. Farmers turn up at a prearranged place with their produce, and customers buy directly from them. Over time a relationship will be built up between producers and consumers. Box or subscription schemes are the classic CSA model, with a range of options for how much each member pays and how much volunteer work is expected of each one. Whole farm CSAs can be entirely consumer driven,

4

Map showing cropping areas at Øverland Gård CSA near Oslo, Norway.

with a group buying land or a farm, and employing a farmer to grow food for them.

Though the schemes are very varied and different, there are enough of them to distinguish certain common features:

- A dedication to quality (organic or biodynamic)
- Diverse, with many crops and sometimes livestock such as chickens, sheep or goats
- Labour intensive
- Multi-dimensional, producing, processing and delivering
- Educational and cultural dimensions
- A dedication to environmental improvement

It has been found that there are many benefits in CSAs. For consumers, and that's most of us who like to eat, we get fresh food from a known source while improving understanding about food production and the real costs involved. We expand our knowledge of new and traditional varieties and become familiar with the produce of different seasons. By having access to a farm as a resource for education, work and leisure, we improve our health through better diet, physical work and socialising. This in turn gives us a sense of belonging to a community. In a wider sense we help to influence the local landscape, and help farmers to use more sustainable methods.

Farmers benefit by receiving a more secure income and a higher and fairer return for products by cutting out the middle man. They can raise working capital and financial support from local communities and become more involved in the local community. A CSA can respond directly to consumer needs, and will encourage communication and cooperation with other farmers.

Society as a whole can benefit by having fewer 'food miles', less packaging and more ecologically sensitive farming with improved animal welfare. Care for local land is encouraged and local economies are enhanced. CSAs improve social networks and social responsibility, and improve the sense of community and trust in an area.

Establishing some form of CSA would be an ideal way for a rural Ecovillage to use its land, to provide meaningful employment for some of its members, and to integrate more fully into the wider area and community.

Velwell Orchard

Velwell Orchard started as a small biodynamic market garden in 1992. It turned out that selling vegetables never made enough money to cover all expenses, not even enough to live on for the owner and his helpers. In 2008 the owner retired, and the small enterprise was taken over by Jeremy Weiss, who had started as a fourteen-year-old volunteer, and was now trying to make this his living. But the market garden was not turning a profit, not even able to give him a modest income. The economics of supply, demand and market competition was not working.

In 2011, realising that a major change was needed, he invited friends, family and volunteers to a harvest celebration and explained his dilemma. It was decided that it would be more appropriate to emulate nature, give away the produce freely to the volunteers who came, and ask them to donate whatever they thought appropriate to the enterprise, not as a barter for the food, but as a gift, freely given.

The enterprise took on a new life, and after a year and a half of the new system there was enough money to keep the finances ticking over, capital had been raised for a new polytunnel greenhouse for raising more vegetables, and volunteers came every Wednesday to help in the work and collect produce. (Read the full article in *New View Magazine,* issue 68, summer 2013.)

Farming in cities

Even though we associate farming with a bucolic country life, cities have always grown a significant amount of food. In an urban setting we are forced to use small-scale, labour intensive, precision gardening. This is both inevitable and appropriate, plots are small, and people are plentiful. Farming will be more designed, and look less like nature, but still based on natural systems and principles. We might use systems that require more technology, that are further away from pristine nature, bearing in mind that the overall aim is to produce food for people living in the city. We might use some of the following techniques.

Hydroponics is the cultivation of plants in a nutrient-rich liquid. There is no soil, the plants send their roots down into a gravel type matrix which is flooded periodically with the liquid.

Aquaponics uses fish and their wastes as the nutrient for plant cultivation. Here the plants grow in the runoff from a fish tank, where the waste materials from the fish constitute the nutrients that the plants require. This is a system that has been practised outdoors in the Far East for many centuries, and requires relatively little modification for a modern set-up. In a city, the system can be modified to fit into greenhouses and other structures.

With **Aeroponics** plants are cultivated with their roots suspended in the air, surrounded by a nutrient rich mist, maintaining a constantly renewed layer of moisture and nutrients on their root surfaces. Again, this can be complex technically, requiring tanks, plumbing and a technical infrastructure, but nutrients can be gained from wastes generated locally, and the produce would have very few food miles.

These systems are all getting us away from nature, and into areas which may have unexpected complications. On the other hand, for cities and urban situations they may well have a place. Tom Levitt reports in *Resurgence and Ecologist* magazine in the autumn of 2013 that one of the world's largest aeroponic farms has been running for more than ten years in Singapore, producing salads and herbs for local supermarkets.

Rooftop gardening is gaining a hold in several cities, and gardening waste areas, unused lots and allotments have been in use for a long time.

It's clear that as more people move into cities, we need to address the challenge of feeding them, using the resources that a city has, and not being too hung up on the 'back to the land' idea. Many people live in cities, they need to be assured of a good and wholesome diet, not necessarily trucked in from distant farms.

For an Ecovillage or an urban 'Transition Town' initiative, focussing on food can be one of the entryways into the local community. Here are some examples of current city farming projects:

- The *Abundance* project in Sheffield, UK, harvests the city's fruit trees, both private and public. The fruit is shared with the tree owners, and surplus is distributed amongst the volunteers and

Community garden in Forres in Scotland.

harvesters, remaining surplus being given to community cafes and refugee centres.

- *Local Greens* in London is a community organic vegetable box scheme that provides UK produced food all year round.
- *Eagle Street Rooftop Farm* in Manhattan, New York City, grows organic vegetables on 550 square metres several storeys above the street. They have an onsite farm market and deliver produce to restaurants in the area by bike. They also run educational and volunteer programmes.
- *Bright Farms* in Brooklyn, New York, design, build and operate greenhouses in order to grow fresh organic food where it is needed. They have been in business since 2006. They have projects in five states, including Washington DC and New York.

4

Organic Bhutan

One country that is aiming for 100% organic agriculture is Bhutan. This small country of only 1.2 million has set itself this target for both philosophical and practical reasons. Largely Buddhist, people are encouraged to respect life and see this as incompatible with pesticide use.

But it's not all plain sailing. There are problems with the erratic weather caused by global climate change, and in some areas both pesticide and fertiliser use has increased in order to cope with this, but the overall aim of the country has not changed. Pema Gyamtsho, Bhutan's minister for agriculture and forests was quoted last year as saying 'Hopefully we can provide solutions. What is at stake is the future. We need governments that can make bold decisions now rather than later. Going organic will take time. We have no set deadline, we will achieve it region by region, crop by crop.' (Reported *in The Guardian Weekly,* March 2013.)

Biodynamic farming

Biodynamic agriculture has as its main aim to work positively upon the earth and humanity's further development. Humanity stands in a responsible relationship to the mineral, plant and animal kingdoms, and must be able to differentiate between the three. Biodynamic practices aim at strengthening social relationships, developing natural and cultural landscape quality, creating food yields and growing healthy food.

The biodynamic farmer will be aware of the relationships between all factors: earth and plants, plants and light, animals and people, farms and society. In biodynamic farming these relationships are strengthened in a positive way. An extra dimension is added by cultivating an awareness of cosmic forces and their relationships to growing plants. Biodynamic farms are regarded as single organisms with many aspects, and planned accordingly.

Biodynamic farming is based on the eight lectures given by Rudolf Steiner in Koberwitz in 1924. For the last ninety years it has been developed and grown to a worldwide, continuously expanding movement.

Biodynamic preparation being mixed prior to spraying.

The aims of biodynamic farming:

- Stop chemicals being transferred to soil, water and air.
- Strengthen the life-giving frameworks for soil organisms.
- Build up the soil's humus content and so bind CO_2.
- Create an agriculture based on natural cycles connecting plants and animals.
- Care for animals and limit human meat consumption.
- Create a plant production that enhances diversity.
- Replace industrial agriculture with social agriculture.
- Produce nutritious food of a high quality for humans and animals.

Working with the soil to grow plants and animals is more than just another human activity. By observing and understanding the natural world, we also do something to our deeper selves. It has been observed time and again that gardening has therapeutic qualities, helping people who have special needs. One of the aims of biodynamic farming is self-development for the farmer, giving him or her an opportunity to grow spiritually by interacting with the spiritual component of the natural world.

The agricultural spectrum

If we regard biodynamic and organic farming, based on Permaculture design, as the most desirable and appropriate to our planet at this time, the other end of the spectrum would be chemical industrial farming as the least desirable. With a planet teeming with people, and with limited energy resources and land, it's obvious that precision agriculture will produce more food per square metre than the rough, large scale, energy-gobbling practices which are today hailed as 'modern, conventional farming'.

There is nothing conventional about a system of farming that began about a hundred years ago, and is now one of the major drivers of climate change, resource depletion and loss of viable farmland. Contrast this with over ten thousand years of organic farming, and the word conventional no longer applies.

Somewhere in the middle of this spectrum lies 'Integrated Pest Management', something that may hold the realistic key to positive change. It is unlikely that we can persuade the chemical industrial farmers to give up their destructive ways and move directly to organic farming, but we may be able to persuade them to use fewer chemicals by introducing natural ways to deal with pests. Using natural chemical substances already found in nature to combat pests will reduce the harm done by synthetic chemicals that find their way into soil, ground water, insects and animals. Many plants and insects contain compounds that are repulsive to other plants and insects. Locating and extracting these compounds has several advantages when they are applied to crops. They are natural and so break down relatively easily in nature. Unlike synthetic chemicals, pests tend not to build a natural resistance to them over time, and so they remain effective without having to step up the dosage. Used in combination with conservation headlands, hedges and agro-forestry, they enhance the natural diversity that keeps every natural ecological system in balance.

In the non-dogmatic thinking that informs Permaculture, it is often more effective to use small and slow solutions, working towards a long term aim. Farmers who use Integrated Pest Management would be much more open to organic practices when they see natural methods reducing their expenditure on harmful chemicals.

A rich cultural life with plays performed by the people who live at Camphill.

Community Profile

Solborg

Camphill Solborg is a small community of about 45 people that was started in 1977 and where I lived with my family for over seven years. We are still good neighbours and still connected to the community in many ways. As in all Camphill villages, a large number of the residents are people with special needs, and the working life is set up so that as many people as possible can participate and contribute.

Solborg has extensive forests that are logged for timber and for firewood, a farm producing milk, meat, eggs and wool, a large vegetable garden and a highly productive herb garden. Being Norway, there is little that can be done on the land between October and March, it's generally frozen and there is little sunshine. Milk is produced all year round however, and meat can be frozen. Vegetables are stored in a frost-free underground cellar, and in most years will last until the spring.

All the farming is done biodynamically, honouring the hidden forces at work in the soil and the plants. Sowing, planting and harvesting is

Working in the herb garden.

carried out in accordance with the stars, following a calendar worked out every year in Denmark. The herb workshop gathers and dries not only produce from the gardens, but also wild flowers and leaves. These are dried and mixed into a large variety of teas, herb salts, tinctures, creams and mixtures. Fruits and berries are gathered in large quantities and stored as juices and jams.

All these products are used within the community, and are available for sale to a wider public, sometimes at farmers' markets in Oslo only an hour away, at local gatherings and fairs, and at a weekly stall by the local Steiner School.

All the 120 or so Camphill villages throughout the world are by definition Ecovillages, their farming is organic or -67, they are fully featured and provide work, culture and self development, they generally practise a form of shared income, and are conscious of their role as agents of social change. The Camphill movement has its own international architectural company and they build using natural and local materials as much as possible. They are inspired by anthroposophy and generally have strong links to Steiner (Waldorf) schools, biodynamic farms and ethical banking systems. The international Camphill movement, with villages in over twenty countries, has regular national, regional and international meetings and conventions.

For more information: www.solborg.camphill.no/

Tips for facilitators and groups

Make your own companion plant groups

Natural ecologies are made up of groups of different plants, a feature that we try to mimic when we design our own food-producing gardens. Different plants utilise available water, sunlight and nutrients in different ways, and by combining various varieties, we can make sure that the ecosystem is efficient and self supporting.

Make cards of different vegetables, cutting and pasting pictures from vegetable seed catalogues. Note the details you have collected for each plant on its card:

- Light and heavy feeders, nitrogen fixers
- Special needs: shade, water, etc.
- Growth habits: tall, short, spreading leaves, tap roots, climbers, etc.
- Which insects are repelled or attracted by each specific plant

Now mix and match the cards to create a number of groups.

4

Compost making on a farm scale.

Chapter 5. Gardening, Soil and Plants

Everyone should know something about soil, a thin layer on only some parts of the dry land of our planet that sustains most of life. Plants grow in this soil, miraculously combining the relatively simple minerals and gases found in the soil and the atmosphere, together with the sunlight available above ground, into more highly complex organic molecules. This is where our food and nourishment comes from.

Had more people been aware of this life support system, surely we would not have gone down the industrial, chemical agricultural road that we have done. Today we are experiencing one of the greatest losses of topsoil and the most drastic incidences ever of bad soil health in our existence on this planet. As designers, we have some work to do!

Soil architecture

I like to talk about the 'Microscopic Architecture of the Soil'. And it really is microscopic. Between the sand, silt and clay particles, composed of minerals, live millions of tiny micro-organisms, and they are busy breaking down organic matter. Like us, they need air and water in order to stay healthy and alive. So the spaces between the mineral particles have to be large enough for air and water to pass through, but not so large that it creates a draught, or allows the water to drain away. Because of this you should try to avoid stepping on the soil: this will keep it from compacting, so that we maintain the open and free flow of air and moisture through the soil. Soil scientists tell us that average industrial chemical farm topsoil contains about 25% water, 25% air, 49% minerals and 1% organic matter.

Unfarmed soils go up to about 8% organic matter, and well farmed organic or biodynamic soil can contain up to 20% organic matter. Whatever the percentage, 1 or 8 or 20, it is this percentage that gives life to the soil, that makes the difference between a dead and dusty mineral mixture and the living organism we call soil. A teaspoon of this soil may contain up to 700 million bacteria, 5,000 species of fungi, 10,000 protozoa and 20–30 different nematodes. And all these are engaged in breaking down complex organic molecules into simple compounds that can be absorbed by the plants growing there.

Soil scientists divide up soil particles by size, the largest being sand, the smallest clay and in between is silt. A balanced soil, which we call loam, should have a good deal of clay, some silt, and a smaller percentage of sand. A soil triangle will tell you exactly what the percentages should be. If you garden regularly, you will get a feel for this. Whenever you can, pick up a handful of soil, and feel it between your fingers. You might also like to smell it. Gradually you'll get to know the texture and smell of good soil from productive organic cultivation. Generally the darker the soil is, the higher the percentage of organic matter.

If we want to take care of our soil, we need to balance out the minerals Nitrogen, Phosphorus and Potassium, making sure that a free flow of air gives access to Hydrogen and Oxygen, that there is plenty of Carbon and that there are enough of the trace elements Sulphur, Calcium, Iron and Magnesium. The acidity, measured as pH, should be between 5.5 and 7, preferably around 6. But none of this is any good without nourishment for all those millions in our teaspoon, and that nourishment consists of organic matter, either decayed or decaying. And the best addition that can be added is well-made compost. Indeed, it seems that for any given problem relating to plant or soil health, the treatment always seems to come down to the simple advice: 'Add more compost!'

In this process of breaking down organic matter, the bacteria will take up mineral ions, forming colloids and polysaccharides containing minerals, which are then available to be taken up by the root hairs of the plants growing there. Decaying alfalfa or oat straw can produce half a ton of polysaccharides in 1,000 square metres within a week. By making and applying compost, we can help in the production of polysaccharides.

Compost

So how do we produce this wonder ingredient, compost? The first thing we need to be aware of is the difference between organic and technical (or industrial) materials. Anything organic will decompose, if given enough time, even bones and teeth, (but these take really a long time, so don't have too many in your compost mixture). We have managed to surround ourselves with artificial (or technical) materials, most of which don't break down very easily. We really need to be clear about this. That nice jumper that is 50% wool and 50% polyester is a garment that cannot be recycled. It will not compost, neither can it be recycled as plastics. Even that harmless padded envelope that you just got in the post with that really good environmental DVD film is a hybrid, with its plastic bubblewrap surrounded by a paper cover. You can't burn it, it doesn't go in the plastic recycling, and neither will it compost. It's just another addition to the landfill, another problem.

5

This is an opportunity to observe your lifestyle, the things you buy and use, and divide them into industrial or technical products that can be recycled, and organic products that can be composted.

To make compost we need to balance the carbon and nitrogen in the mixture. We want about 20–30 parts carbon to 1 part nitrogen, and plenty of air. Too much nitrogen and/or too little air (anaerobic conditions) will lead to a build up of ammonia. Avoid this, because it smells bad, and doesn't produce good compost. If you're not sure if it's nitrogen or carbon, decide if you would be happy jumping into it and bouncing. If it's leaves, hay, straw or paper you would probably be quite happy rolling in it, and it will almost certainly be composed largely of carbon. If it consists of slimy, rotting grass cuttings, fresh kitchen waste, humanure or animal dung, you would probably not want to get in, and you can be certain that it will contain a good deal of nitrogen.

The science and art of composting is a relatively recent field of study, though the process is based on what happens naturally when organic matter decays back into soil, a process that has been used by farmers for hundreds if not thousands of years. To observe this process let us build a compost heap, laying down brushwood on the ground, to allow air to enter from below, adding alternating layers of kitchen waste, straw, animal dung and leaves in the ratio described, making sure that air can circulate freely up through the heap. The best way to ensure this is to drive some fencing stakes into the heap as you build, and pulling them out after the heap is finished. As soon as the heap is finished we will be able to observe four clear stages:

1. *Mesophilic phase.* This is the first phase and temperatures can reach 44°C. Large numbers of bacteria are present, including E. coli. This takes only a few days, and the heap will sink visibly.
2. *Thermophilic phase.* Thermophilic microorganisms proliferate and boost temperatures up to 70°C. (Not really desirable) This stage is fast and brief, and may be localised in only one area. This does not digest the coarser material. Again, this is a brief phase, taking only days, and the heap will continue to sink.
3. *Cooling phase.* Lower temperature microorganisms migrate back into the pile and start digesting the coarser material. Fungi, mycelia and macro-organisms such as earthworms proliferate. This takes several weeks, or, if it's a cold climate in winter, maybe months.
4. *Curing phase.* A long and important phase, that can last up to a year, and which adds a safety net for pathogen destruction. In Norway, with several months of subzero temperatures, no human pathogens can last more than two years in the soil. Immature or uncured compost can produce phytotoxins that rob the soil of oxygen and nitrogen and can contain high acid levels.

Impatience often results in not giving the compost enough time to 'cure'. Commercial compost production can be very prone to this. The curing will be faster if the ingredients are carefully selected and broken up by a machine into smaller particles. For a domestic situation, a smallholding or farm, individual solutions should be worked out, taking account of the particular conditions specific to the lifestyle of the people involved. In our garden, we are now using a two-year cycle, with the whole of the second year given over to undisturbed 'curing'. Our compost consists of a relatively small amount of humanure, in addition to all our kitchen waste, and the leftover stems and leaves from the garden. Every deposit of whatever material is covered by a layer of half rotten straw, ensuring plenty of carbon.

When planning domestic or neighbourhood composting systems, there are two main factors that need to be considered. The first is that every situation is unique, and it's best to work from the composting principles outlined above. The second is that it really doesn't matter how long the process takes. Once throughput is achieved, what goes in at one end will come out of the other. I would like to propose a 'Slow Composting Movement' in order to produce high quality compost and show solidarity with all the other 'Slow' movements. Many of the problems we are addressing stem from sacrificing quality for speed, and in the new paradigm we need to reverse that equation.

5

Worms can turn your organic waste into a high quality fertiliser.

Mulch gardening and 'Hügelbeds' are variations on the same theme, building further on the principles of composting. In natural forest ecosystems we see how soil is created by layers of leaves and plant wastes gradually accumulating on the surface and slowly decomposing. In mulch

Adding compost to your soil is the key to successful gardening.

gardening we mimic this by adding organic waste material on the top of the soil, rather than by digging up the soil itself. In some places this is called 'Lasagne Gardening', and the layers are carefully arranged to create thin layers of compost between the soil and the mulch. Here the composting process takes place, releasing nutrients for the plants. An additional advantage of mulching is that it deters weeds from growing, thus saving us work.

Back in the forest again, we can observe fallen trees that are slowly rotting, creating patches of nutrient-rich soil supporting colonies of plants. If we looked closely, we would see fungi, mycelium, and microorganisms breaking down the woody material into nutrients for the plants. We can create a similar system in our own garden by digging a trench or a hole, filling it with dead wood like the prunings from our orchard, and then covering the pile with the soil that was dug out. We will end up with a mound of earth. Inside this mound will be woody material that will gradually decay and release nutrients for a long period. When the mound is planted thickly with vegetables, their roots will seek out these nutrients. The mound will gradually settle over the years, and will nourish plants for all of that time. These are called 'Hügelbeds' and large ones can be built using a digger, big logs can be placed in the trenches, releasing nutrients for decades as they decay.

Biochar

Amazonian Dark Earth is a technique practised by indigenous Americans up to the arrival of the Europeans, which then fell into obscurity, only to be taken up again by, amongst others, Wim Sombroek in the 1990s. Biochar strategies entail making charcoal, which is then added to the soil, creating spaces and soil microclimates that encourage the development of soil microorganisms. This is best covered in Albert Bates' book *The Biochar Solution: Carbon Farming and Climate Change.*

Nitrogen fixing in organic agriculture

Nitrogen is a key element in the nutrient cycle. One family of plants, called legumes, have the ability to fix nitrogen directly from the air. This is done by the bacteria associated with specific plant roots. These are called rhizomes and live in a symbiotic relationship with the legumes. The bacteria get some of their energy, water and minerals from the plants, in return giving the plants a rich supply of nitrogen. Some rhizomes are specific to certain plants, while others are more generally associated with several species. The nitrogen enters the plants via fine root hairs, stimulating cell growth of nitrogen knobs on the actual root fibres. Nitrogen molecules are split into ammonium, which requires a high energy use by the bacteria they get by burning carbohydrates created by photosynthesis. To optimise the process it is important that the plants have good light, plenty of water (but not too much) and air. The optimal temperature for the process is 20–25 degrees Celsius, but measurable nitrogen fixing has been measured at even zero degrees. Trace elements such as iron, molybdenum, sulphur, cobalt, copper, calcium, and magnesium are important. Too much nitrate hinders the process, so using fertilisers with artificial, chemical nitrogen is harmful.

When combining legumes with corns and grasses, the non-legumes will grow healthier on account of more nitrogen being available from the leguminous root systems.

Rhizome bacteria can be used to enrich poorer soils. This is very

5

relevant when introducing new plants, when cultivating new ground or improving soil that has been waterlogged and drained. Soil can be taken from fields that are rich in legumes and then spread over the new soil. The rhizomes are not tolerant of light, so this is best done in the evening or in overcast weather, and the new soil worked in straight away. You can also use bacteria rich water to infect the seeds when you sow legumes.

Common nitrogen fixers are:

- Lucerne grass (Medicago)
- Clover (Trifolium)
- Beans (Phasoleus)
- Lentils (Lens)
- Lupins (Lupinus)
- Soya beans (Glycine)
- Bird's-foot trefoil (Lotus)
- Peas (Pisum)
- Vetch (Vicia)

What plants do with soil

The role of soil microorganisms to break down organic matter is only one half of the amazing process that supports life. Plants, utilising the nutrients that we have seen are released by this process, build up complex organic molecules. All green plants use a process called photosynthesis to achieve this. Photosynthesis can briefly be described as: Water + carbon dioxide + light = sugar + oxygen. In chemical language it looks like this:

$$6CO_2 + 6H_2O + \text{light (photosynthesis)} = C_6H_{12}O_6 + 6O_2$$
$$\text{Carbon dioxide} + \text{water} = \text{energy} = \text{glucose} + \text{oxygen}$$

What we find here is a nutrient cycle. Organic matter is broken down in the soil by a succession of bacteria, producing simple nutrients that are then available for plants, which build up complex nutrients that animals and humans can consume. Sunlight drives this cycle, minerals, gases and water fuel it, and the bacteria maintain it. It is self repairing and as long as organic matter is added, it will develop and grow in mass and complexity. Our task, as gardeners and farmers, is to encourage this cycle, and design systems that will ensure an abundance of yields, which we can then use as food or raw materials. The key ingredient is soil. If the soil is living and healthy, the plants will grow well, and there will be abundance for us and for future generations. If we neglect the soil, the plants will be unhealthy, and unable to produce high quality products over time.

Plants have many components: roots, root hairs, root nodules, stem, leaves, flower, fruit and seeds. Plants also go through a life cycle that we need to be aware of: seed, sprout, cotelydon leaf, root, stem, leaf, flower, fruit and back to seed. Regular work in a garden will teach anyone these processes through experience, which is the best teacher. One way of enhancing the learning process is to draw the plants at various stages in their cycle. In our technological age many people will be tempted to get there faster and more accurately by photographing, but this is counter productive. The act of drawing involves close observation, which is usually not the case when snapping a picture. Drawing the same plant, day after day or week after week, will teach us much more.

Plants also have different contexts and different needs: soil, water, air, gases, minerals, mycelia, microorganisms, insects, animals and humans. Again, observation is the key, noting where and when plants grow well or badly, and how changing the amount of water, the shade, and the care will affect the plant.

In addition to the physical and measurable components already mentioned, there is the cosmic influence of the sun, the daily cycle of light and dark, and we can easily observe how some flowers open and close their petals in response to this. Many plants grow towards a light source if put in a dark place. Most experienced gardeners that I have talked to have also observed the difference in plant growth rates with the waxing and waning of the moon. Closer and more subtle observation will reveal influences caused by the position of the planets and heavenly constellations, and by the thoughts, feelings and emotions of people who work with plants. This is recognised in our traditions when we refer to some people as having 'green fingers', people who seem to be able to

5

Gardening in pallet boxes can be done almost anywhere.

grow plants easily and well. Systematic and scientific research has been done by biodynamic farmers and others for nearly close to a century, and we will look more closely at this later.

Plant genetics – breeding or modification?

Conventional archaeology tells us that we have been farming for ten thousand years, roughly since the so-called 'Neolithic Revolution'. Bill Mollison studied the Australian Aborigines and the way they lived off the land, and came to the conclusion that they had been modifying both plants and their environments in order to create greater yields, to such an extent that their practices could be called 'agricultural'. We now suspect that farming is much older than ten thousand years, and that human beings have understood a great deal about plants, and made many interventions and modifications to them for many thousands of years.

One of these plant modifications you can do in your own garden.

Plant a crop of organic peas, select the best plants, save the seeds, plant them next year and do the same the year after. Within a few years you will have a local subvariety, better suited to the exact soil and climate of your garden, and yielding slightly better than the original seeds in the packet that you bought. Farmers have done this for thousands of years, creating local varieties with a wide range of intended characteristics. This is part of a long tradition of farming and gardening, a combination of building organic soil and human ingenuity. Advanced plant scientists can also breed seed selectively, cross-pollinating to enhance desired characteristics. This has also been done for thousands of years. If you buy commercial seeds in packets, check if they are F1 Hybrids. These generally produce sterile seeds, and there is no point in growing them for seed. Most organic seeds are not F1 Hybrids, and if you seed swap with gardeners you know, you will have no problems.

Almost everyone who can grow a few plants can select seeds. Plant breeding can be practised by those who study plants, seeds and pollination patterns. Most rural Ecovillages should be able to build up a selection of staple crops based on saving their seed within a few years. It would make sense to invest in people and infrastructure to create a stable and cumulative seed bank. In some countries it is illegal to save some seeds, and in most it is now illegal to sell your own seed unless it is registered. Seed swapping is still mostly legal, and doing so on a small scale within known groups such as Permaculture Associations, organic gardening groups or allotment holders you should have nothing to worry about.

What is worrying is that international companies are buying up seed rights, patenting varieties, and making it illegal for people to grow plants that were part of our common heritage. A close watch needs to be kept on seed rights, and the national and international agreements regulating these rights. Luckily, 'food security' and 'food poverty' are now increasingly being treated seriously, and it is clear to everyone in these fields that if ownership of our seeds passes on to multinational companies whose main aim is profit for their shareholders, we may go into deep insecurity and poverty when it comes to food.

During the last few decades the development of genetic engineering has given us the opportunity to genetically modify (GM) certain varieties. This was originally hailed as a miracle of modern science and claimed to be able to solve food crises, global climate change, soil erosion, salt tolerance, weeds and pests. This has not lived up to its promise.

In a report published in 2011, it was claimed that in the twenty years since GM crops have been in use, hunger has reached epic proportions, and the use of synthetic chemicals has been on the increase, despite biotech companies claiming to reduce such usage. Soya growers in Argentina and Brazil are using twice as much herbicide on their GM crops as on their non-GM crops. In India, a survey by Navdanya International showed that pesticide use rose 13-fold after GM cotton was introduced. (*The Guardian Weekly*, October 2011.)

I would suggest that Ecovillagers, Permaculture gardeners and growers stay clear of GM crops. The future belongs to heritage species, local seed saving and organic soils.

Plant communities

Observing natural ecologies we find that plants usually grow in groups, and that there are groupings that reoccur. Traditional agriculturalists discovered this long ago, and sometimes these groupings took on additional, cultural and spiritual qualities. The Jewish tradition of the seven species of wheat, barley, grapes, olives, figs, pomegranates and dates is mentioned in the Bible, and forms part of the celebrations of many Jewish Festivals.

In North America another grouping developed that is often referred to as the 'Three Sisters'. Corn (maize) grows on tall stems and gives an opportunity for beans to climb up on these stems. Neither of these plants uses a lot of ground space, which is then covered by squash spreading its leaves and fruit below the other two. In addition, these three give a tasty and balanced diet.

It's well known that other combinations have different advantages. Carrots deter the onion fly, and onions deter the carrot fly; planting them together is a good plan. Marigolds (calendula) are highly unattractive to nematodes, which damage roots and stems. Planting marigolds around the vegetable garden looks nice, their petals are tasty in salads, and they give a useful protection.

Dome greenhouse in Arctic Norway.

You can design your own groups of plants by mixing and matching their habits and requirements. Make a list of the vegetables that you want to grow and note which are light and which are heavy feeders, which of them fix nitrogen, and any special needs they have, shade, water, and so on. Then make notes about their growth habits, tall, short, spreading leaves, tap roots, climbers, and so on. Now find information about which insects are repelled or attracted by each specific plant. Now you are ready to form the plants into groupings, making sure that you have a good balance between the various needs of the different plants.

Food forests and forest gardens

Building on the principles of companion planting, we can create more complex 'guilds' by weaving together plants of different sizes, aiming to achieve something that nature creates when it allows ecologies to grow to maturity, the best example being the savannah and the temperate forest. Here not only do we have a mix of different plants on the ground, but we take size and height with us, and include trees that are several metres tall.

Trees have a number of highly useful functions. They hold water, they clean the air, they stabilise soil and rocky slopes, and they provide shade and wind breaks. A forest is a single organism, a synergy of multiple units. It's highly dynamic: condensation, evaporation, photosynthesis, respiration and transpiration all occur at a high rate. As the forest establishes itself through succession it moves from being a young forest with high input and low yields to a mature or old forest needing only low inputs and giving high yields.

We have covered the idea of food forests in more detail in the last chapter.

Rotation

In our small domestic vegetable garden at home we have divided our vegetables into four main groups: Potatoes, Legumes, Brassicas and Roots. These are not rigid definitions, but the main things that we grow in four areas which we rotate every year, giving the soil a change in emphasis from the heavy feeders like potatoes to the nitrogen-fixing legumes. In addition to improving soil quality, this also helps to break the build-up of harmful pests that attack brassicas, carrots and other plants. Some things go anywhere at all times. Garlic and marigolds might be found in all the beds. Lettuces and radishes, being fast growing and quickly harvested, might be put in anywhere there is room.

This system is known as rotation, and has been a well-established farming practice for hundreds of years, with many variations. Some farmers have developed highly sophisticated rotations lasting up to seven years, working with natural rhythms to deter pests and build soil quality. In one garden we worked on in Lincolnshire we included chickens in our rotation. The vegetable beds were arranged around a central chicken run, and with a well planned system of fencing and gates. We could allow the chickens access to any bed we wanted to at any time, to clean up after a harvest, and manure the soil with their droppings. When they had done their work, we closed the bed off again and replanted. Another strategy is to have a 'Chicken Tractor', a movable chicken run that can be placed over a given bed at a given time, allowing the chickens to do the same job.

Chicken tractor.

In order to keep a strong learning curve we initiated a garden schedule that we updated every year, taking into account the annual variations of late and early frosts, rainfall and temperatures. In this way we built up a very useful schedule of tasks for each month.

Garden Schedule

	January	February	March	April	May
Early Pots			Chit	Plant	Earth up
Main Pots			Chit	Plant	Earth up
Broad Beans			Sow 1	Sow 2	Pinch tops
Leeks				Sow seeds	
Early Peas			Sow 1	Sow 2	Sow 3
Fresh Peas				Sow 1	Sow 2
Dried Peas				Sow	
French Beans					Sow 1
Swiss Chard				Sow 1	Sow 2
Courgettes				Sow indoors	
Runner Beans					
Kale				Sow	Thin out
White & Red Cabb				Sow	Thin out
Savoy				Sow	Thin out
Late Broccoli					
Cauliflower & Sprouts				Sow	Thin out
Chin Cabb					
Turnips				Sow 1	Sow 2
Tomatoes			Sow indoors		Plant out
Early Broccoli					Sow
Onion sets & Shallots			Set		
Carrots & Beets				Sow 1	Sow 2
Swedes & Parsnips				Sow	Hoe and thin
Radishes				Sow and pick all the time	
Lettuces					Sow and pick all the time
Garlic			Set		
Flowers			Sow whenever		

Adjusted for temperate climate. Lincolnshire Wolds in England, c. 130 metres above sea level, 500–600 mm. annual precipitation, 53 degrees N.

June	July	August	September	October	November	December
	When flowers open, Dig up Earth Up Pick	Pick	Dig up	Dig up		
Sow 3	Plant out into Early Pot space Pick Sow 4	Pick onwards			Earth up	Pick
Sow 2 Pick	Harvest when pods split Sow 3 Sow 3	Pick often Sow 4				
Plant out Sow		Pick often Pick often		Pick onwards		
Sow	Pick to thin	Plant out	Pick onwards	Pick onwards		
Sow 3	Sow Pick 1	Thin out Pick 2	Pick 3		Pick onwards	
Stake	Mulch Plant out Loosen roots	Dig up	Pick Dry in sun			
Sow 3	Pick onwards		Harvest			
			Harvest			

Record-keeping is part of the important Permaculture principle of observation. It also fits in nicely with our seasons here in Scandinavia. During the winter, snow blankets our garden for many months, so it's a welcome break from the physical work of gardening, and checking out last season's patterns gives us ideas for the next summer, which we then can plan in.

Garden design

One of my favourite definitions of Permaculture is that it is a series of questions leading us into the future. Taking that as our starting point, maybe we can set some questions about the garden, and see where the answers lead us. I found the following questions in *Smart Permaculture Design* written by Jenny Allen:

- How much food and produce are you aiming for?
- How much time and energy do you have?
- What plants and structures do you want?
- What are your soil variations in different areas?
- How can you make the garden low maintenance?
- What beneficial micro-climates can you create?
- Is there a view to enhance or block out?
- Is there a frost problem? If so how can you reduce it?
- What does each plant require?
- How can you optimise water use?
- How will the garden evolve over time?

These questions might form the basis for a meeting of the people involved in growing food for an Ecovillage group. Answering them may well go a long way to designing the farming or gardening systems that you want to use. They may also help you to design the physical layout of the gardens.

Mandala garden layout.

Student campus at Lotan. Photo: Sarah Bang

Community profile

Lotan

Kibbutz Lotan was founded in 1984 in an area of the Arava Desert in southern Israel that contained a dozen kibbutz communities, democratically run, where there were no private incomes and all the means of production owned by the community. This Southern Arava Regional Council is possibly unique in the world, containing only free communist style villages. Of these villages, three have been cutting edge ecological communities. Kibbutz Samar was the first to explore large scale solar power generation, Kibbutz Ketura founded a university institute for graduating students in environmental studies, and has a large research station devoted to productive desert plants. Kibbutz Lotan established a Permaculture educational facility, linked to Living Routes, which ran a three-year degree course set entirely in Ecovillages around the world.

I have worked with Kibbutz Lotan for twenty years, teaching Permaculture and Ecovillage design at their facility, and watching their Green Campus grow and develop to be one of the finest of its kind. Their extensive gardens contain really good teaching facilities for soil, compost and plant studies, and the produce grown there is to be found at nearly every meal in their communal dining hall.

The outdoor washrooms at the Lotan Campus where students brush their teeth, built out of mud and waste materials.

Spiral garden built by students as part of their Permaculture Design Course.

Being set in an extreme desert environment means that building up soil has been of paramount importance. Good regular mulching, a choice of drought-tolerant plants, and finely tuned drip irrigation has created a rich and healthy soil. Students are taught soil science and plant botany, and there are several composting systems at work.

The Green Campus, built almost entirely of local mud and earth, with straw bale insulation to keep out the fierce desert heat, closes the circle by teaching cooking and food preparation using solar energy and clay ovens.

For more information: http://www.kibbutzlotan.com/center-for-creative-ecology/c8ue

Tips for facilitators and groups

Backcasting

This exercise I came across in the Transition Town literature. It involves visioning a picture of your community twenty or thirty years ahead. This may entail a session with small group discussions, each group taking on a particular aspect, and then sharing these aspects in a plenum session.

Once the picture has been built up and described, and everyone involved feels that they have a mental picture, and there is a set of keywords and descriptions written for everyone to see, a timeline is created, stretching along a wall. You can write the years in to create a clear timeline from today to the future community.

Ask the small discussion groups to think back over their particular contribution, and to mark in, with Post-it notes, what has to be put in place when for their contribution to happen. For example, if it's a school, when does it have to registered with the authorities? For when does it have to be built? When does the application for building permission have to be sent in? If it's a Permaculture teaching and research centre, how long does it take to get established, and what kinds of steps have to be taken along the way? The timeline can be as general or as detailed as the time allows and the participants want.

Working backwards from a desired result can give the group a pretty good detailed plan of action.

Backcasting presentation from Applied Ecovillage Living course held at Findhorn in 2014.

South facing greenhouses capture solar energy at Munksøegård Ecovillage in Denmark.

Chapter 6. House Design

The way we build has an enormous impact upon our environment. It has been estimated that 47% of total global energy use goes into building. This includes the production and transport of materials, the actual construction process, and maintenance, including heating and cooling through the building's lifetime. Clearly this is an area where we can make an enormous difference to the human-induced climate change effects, and in reducing energy use and pollution generally.

This chapter will deal with building design strategies and techniques, while the next chapter will look more closely at the actual materials we use in building.

Design strategies and techniques

Design is fitting the building into its surroundings. Climate is probably the most important context, and anyone contemplating 'new build' is advised to spend at least a year observing the site, to collect as much material as possible from weather data records, and to talk to older people living in the area about unusual weather events they have experienced in their lifetime.

The obvious things to look for are the directions that the sun shines on to the site, bearing in mind the differences between winter and summer. The further north you go, the larger the difference, and sheltered south-facing alcoves can be very comfortable to be in even quite early in the year. Summer sunshine can be much too hot to endure comfortably, so the planning of outside areas with shade should be factored in, especially in southern countries.

The angle of the sun in midwinter and midsummer can be used to great effect. A simple overhang above the windows can be very effective in shading from the high midsummer sun, and let the much lower midwinter sun flood into the house, heating up dark massive materials such as stone, concrete or water-filled containers, which will keep the heat for many hours. The same heat sinks will stay cool in the summer, and so help to keep a comfortable indoor temperature.

The direction of prevailing winds should be noted, and here it's really useful to talk to people who have lived there many years. In areas prone to forest or grass fires, it is also essential to note from which direction these might come. Humidity is another feature that should be taken into account. An area of high humidity might not benefit from a multitude of streams and ponds, whereas a dry place may be made a lot more comfortable if running water is planned in.

Only careful observation, preferably noted down in a log or diary, will help to identify these features.

Local building materials can also be noted. If your site is in dense, mature woodland, timber may be the material of your choice. Stone, earth, clay and sand can all be used in building, and if they are right there on your site, you may be able to make significant savings in transport. Derelict houses on site or in the vicinity may be sources of reusable building materials.

Energy consumption of buildings is a feature that may easily be overlooked at the design stage. Careful planning of spaces, incorporating the information obtained by thoughtful observation, can significantly reduce the energy needed to heat and cool a building.

In addition to the physical particularities of the specific site, there are social and financial considerations. Obviously there will be a completely different building if it is intended to be used by one person rather than a family of five, and if you are an Ecovillage group, you may well consider the beneficial effects of several people or groups living in one large house, or a connected row of terraced houses. You may also consider the changes that a family goes through over time. Typically a family starts off as a couple, they may have one or more children, requiring a good deal more room, but then eventually the children will grow up and move away, leaving the couple back as a twosome. But then in later years there may be extended visits from children and grandchildren, again needing room for both staying and playing. Some communal groups solve this by building relatively small family houses, and having separate guest rooms available for anyone coming to stay, and extensive communal play areas for children.

In the early years of kibbutz, children slept in children's houses, the laundry was communal, and everyone ate all their meals in the dining hall, reducing family homes to the absolute minimum, often just one or two rooms, even toilets and showers were communal. Today, co-housing projects often have shared laundry, shared workshops for bicycles and garden tools, play areas for children and a common house that members eat some of their meals, and which may even contain a communal library.

Other very important factors are time and money. Generally speaking there is a trade off between the two; if you have lots of money but very little time, you will find yourself hiring professionals to do most of the work, from designing and planning through to the actual work itself. If the opposite is the case, very little money, but lots of time, it's obvious that the best solution is to do as much work as possible yourself. If you have neither time nor money, you will find yourself seriously challenged, but if you have lots of both, you can really enjoy yourself! This is as true for a group as it is for an individual person. However, the synergy multiplier effect of being a group can really make a difference. There is also a strong likelihood that within a group there will be a range of relevant skills, which could make the whole process much more enjoyable and productive.

To sum up, we need to know our context before we can start the design process:

- The general climatic and environmental context
- Wind directions by season
- Sun angle, both vertical and horizontal
- Annual temperature fluctuations
- Existing plants, from the smallest herbs to the largest trees
- Animals, wild and domestic
- Existing buildings
- Locally available materials
- Social and financial contexts

Permaculture zoning

This is one of the classic tools of Permaculture Design, and I would urge anyone on the designer's path to become well acquainted with it. It's generally easier to begin by applying it to a smallholding situation until you feel really familiar with the different zones. A group designing a community might then start applying it to a much larger context, and create multi-layered zoning with complex sectors and vectors. This would be an extremely useful exercise to practise designing the actual community plan.

When you become really familiar with this tool, a great deal of fun can be had by scaling up and redefining the zones to apply it to regions. Even more fun by scaling it right down to help plan your kitchen or bedroom layout.

Zone 1. *Homes and food security gardens.* This includes the house itself (seen as Zone 0) and workshops that are in daily use, as well as gardens for herbs and intensive vegetables, especially those that can't be stored. Trees would be limited to those giving shade for the houses, or ones that give regular and frequently used fruits.

Zone 2. *Close spaces and orchards.* This is still an intensively maintained area, any irrigation that needs to be controlled, egg collecting and daily milking can be located here. The actual sheds for these might be on the border between Zone 2 and 3 giving access for the animals to range in the next Zone. This area could also include pruned trees and mulch gardens.

Zone 3. *Larger open spaces and gardens.* Water would be available in ponds and streams, but otherwise unmanaged. Trees might not need regular pruning, but could give yields in the shape of annual gatherings of nuts and fruits. Animals would be able to range freely.

Zone 4. *Reserves, fuel forests, windbreaks, and so on.* The managed, designed space that we surround ourselves with gradually gives way here to the semi-wild. From this area we gather from the wild, typically for firewood and timber, and for those who eat meat, this would be where we might hunt small animals. Mushrooms and berries gathered here would be a welcome addition to your food.

Zone 5. *Wildlife corridors, native plant sanctuaries.* We stop designing, and let nature take over freely. We need to see nature in action, partly to see how things operate, and partly to nurture our own spirits by connecting to the natural world in a quiet and meditative space.

These zones are in theory concentric circles, but need modifying with sectors and vectors.

Sectors are parts of these circles having specific attributes, such as the sunshine generally coming from a southern direction in the northern hemisphere, the prevailing wind direction, and fire risk areas adjacent to the site. Occasionally Zone 4 or 5 might be brought quite close to the house in a wedge, giving wildlife the opportunity to move across the site, or the location of a barn, shed or workshop might extend Zone 1 or 2 further away than the perfect concentric circle. The actual topography must be taken into account, and should play a large part in determining what goes where.

Vectors further modify the pattern by creating dynamic flows cutting through the site, such as watercourses, existing or planned roads and tracks, wildlife corridors and the hills and slopes of the site. While Sectors can be regarded as direction-specific, Vectors are often more serendipitous and move more or less where they want.

In the planning of any specific site, these three ideas are further modified by the elements that we find there: existing buildings, roads, streams, hills and forests. We say that Permaculturalists work with what they get. Sometimes there is a need to make major landscape changes, but generally it is better to make do, and not move hills or major watercourses. This means that each design will end up as a unique exercise, applying the theory outlined here to a specific location. It might be said that understanding the theory is the Science of Permaculture, while applying it in practice is the Art.

In recent years Permaculturalists on the cutting edge have looked more closely within Zone 1, and seen Zone 0 as the house. This has been further refined by going deeper into Zone 00, where we find our psychological and/or spiritual aspects. We may want to dig further into this idea by asking ourselves a series of questions:

6

- Have we set aside a space for spiritual practice?
- Is this a purely private matter, or may we as a group want to have a space for meditation or similar practice?
- How does the layout of the building affect us spiritually?
- Is there a natural progression through the rooms or spaces?
- Do we want to soften the harshness of right angles and straight lines by putting in more angles and curves?
- How do we look after ourselves as a group, care for each other, show respect and solve conflicts?
- Does this need to be reflected in our buildings?

Site planning

Any detailed design of a building has to include a careful consideration of the actual site itself, and how it will be used while building is taking place. Here is a summing up and a few considerations:

- Renovate if possible. This may go against the advice of modern builders, but they are generally obsessed with time (time is money) and the use of prefabricated industrial materials that may not fit into the dimensions of an older building.
- Evaluate site materials. Here the list you made while observing the site will be of great use. Consider transportation. Move things as little as possible, and when you do, put them in the most effective place.
- Locate building to minimise environmental damage. One good Permaculture principle to bear in mind here is that you should leave those parts of the site that are the most beautiful or the best, as they are, and work on the least attractive areas, making them more attractive.
- Consider solar energy in siting and landscape. Again, your observations are the key here. Energy efficiency is a top priority.
- Consider existing landscaping. Creative siting within small landscape variations can give the buildings a great deal of

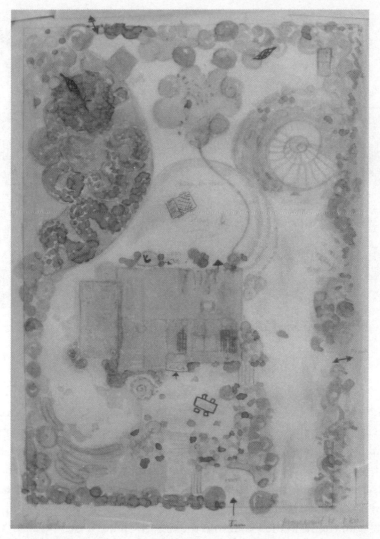

Site plan presented at a design course in Iceland.

character. Try to avoid bulldozing the whole site flat. Design your landscape holistically, considering energy and food.

- Optimise use of materials, and design for recycling and future reuse of materials.
- Consider grey water recycling and use.
- Avoid health hazards such as radon and electromagnetic fields.
- Smaller is better.

Solar energy being captured at Dyssekilde Ecovillage in Denmark.

Energy and heating

In the northern hemisphere, as the sun moves across the sky from east to west during the course of the day, it will shine on different parts of buildings. In your house where you live, you might typically want it to wake you up in the morning, give a good strong light to the kitchen and to the main living room during the day, and warm your terrace in the evening. So you would design your house accordingly, bedrooms on the east, main rooms to the south, and the terrace on the west.

In a climate with hot summers and cool winters, an overhang would shelter the rooms inside from direct exposure to the high summer sun, and give access to the warm winter sun, much lower in the sky, which would flood the rooms with heat.

The sun, shining through glass on to a dark surface, will heat up the surface, which will in turn heat up the air contained within the glass. This is the classic greenhouse effect, and this effect can be used to heat buildings. There are three basic ways of approaching this:

Solar gain through ceiling windows helps to heat this communal building at Munksøegård Ecovillage in Denmark.

6

1. Direct heating allows the sun to shine in through the window and heat up floor and walls that in their turn will heat the room. If these are dark coloured and made of massive materials, they will store the heat, releasing it slowly overnight.

2. Indirect heating allows the sun to shine on to a darkened wall which in turn radiates heat into the building, or which circulates warm air into the room behind it. This can be done by having a vented opening at the top and bottom of the wall, opening into the room. Hot air, warmed by the wall, rises and enters the room, sucking cool air into the space from the room at the bottom. A circulation is established and the room is warmed. This wall is often called a 'Trombe' wall.

3. Isolated systems consist of collectors placed on the roof, on walls, or even adjacent to the building. These collectors warm up air, water or other substances that transport the heat into the building, releasing it where it is needed. The heat could

be stored in an insulated water tank, or a massive rock storage cellar. In some designs, this can be heated during the summer, taking several months to warm up, and the heat released during the winter.

These three systems are not exclusive, and good design would combine all three in a single building.

Passive and active houses

A generation ago there were very few architects and builders who approached design with a strong environmental awareness. Industrialisation was the buzz word, materials were often hazardous and aged badly, and it was thought the same design and the same materials could be used from Iceland to Sri Lanka.

This has changed over the past few years, and now there is some serious work going on in house design, with official standards beginning to be applied. One of these is the Passive House idea.

For a building to be considered a Passive House, it must meet criteria that are achieved through intelligent design and implementation of the five Passive House principles:

- Thermal bridge free design
- Superior windows
- Ventilation with heat recovery
- Quality insulation
- Airtight construction

In addition there are energy demands:

- Heating energy demand not exceeding 15 kWh per square metre of net living space
- Total energy must not exceed 120 kWh per square metre of treated floor area per year
- Maximum of 0.6 air changes per hour

Years ago we used to joke that *passive* houses needed *active* people to run around opening and closing windows in order to regulate the temperature according to sun and wind. This is now no longer the case, with thermostat-controlled heating and ventilation, this can be built in automatically. Of course, the house will become highly technical, and the initial investment rises sharply. Maybe it's cheaper to have those active people!

Passive house in the first housing cluster at Hurdal Ecovillage in Norway.

Since then the idea of the Active House has been defined, rendering the joke even more obsolete. This is a vision of buildings that create healthier and more comfortable lives for their occupants without impacting negatively on the environment.

According to their website, the Active House idea is as follows:

The Active House vision defines highly ambitious long-term goals
for future building. The purpose of the vision is to unite interested
parties based on a balanced and holistic approach to building design
and performance, and to facilitate cooperation on such activities
as building projects, product development, research initiatives and
performance targets that can move us further towards the vision.

The Active House principles propose a target framework for
how to design and renovate buildings that contribute positively
to human health and wellbeing by focussing on the indoor and
outdoor environment and the use of renewable energy. An Active
House is evaluated on the basis of the interaction between energy
consumption, indoor climate conditions and impact on the
environment.

Design values

Design should reflect spiritual, social, economic and ecological values.
Buildings have been called our third skin, after our clothes (second skin)
and our actual skin (first skin). Just as we choose our clothes not only to
protect us from the elements, but also to make a statement or to make us
feel spiritually or culturally comfortable, we might consider our buildings
in the same way.

We are what we live in. When we plan our buildings, we are also
planning what kind of fellowship or society we want to create. A house
designed for a dozen residents does not have to be much larger than many
large family houses in the affluent west today. Without a clear idea of how
we want to live together, it's impossible to plan and design buildings.
When we inherit old buildings, we are also inheriting a social pattern, and
need to be very aware of this. Do we need to rehabilitate the buildings to
reflect the new kind of social system we are creating?

As we saw above, designers should ask site-specific questions. Where
are the sun and the wind? Do we want sun in the dining room and kitchen
for the morning time? Perhaps we orient the living room towards the west,
so we can gather in the evenings and watch the sunset. Storage of food and
fuel can be along the north wall (in the northern hemisphere). Food stays
cool, and the woodpile will be protected from cold north winds. Large
south-facing windows let in the winter sun (in the northern hemisphere).
An overhang gives us a place to sit and shade from the hot summer sun.
This is design in action, working with the realities of the environment and
the needs of the human being.

We make the buildings, and the buildings make us!

Forgebank Co-housing community members in 2012.
Photo: http://www.lancastercohousing.org.uk

Community profile

Lancaster Co-housing, Forgebank

The Lancaster Co-housing (LCH) group was formed several years ago in order to create a community based on cooperation, and with a clear boundary between the private and the common. In the aftermath of the financial crash in 2008 property prices fell and many development companies went bankrupt. It was a buyer's market, and LCH found an old derelict industrial property at Forgebank just three miles out of Lancaster on the banks of the river Lune, in the village of Halton.

The group bought the property for £600,000 in 2009 and already had a good deal of the organisational infrastructure in place, including architects, planners and lawyers. The total investment for planning and building the forty terraced houses and installing the heating, sewage, energy and water systems came to around £11 million. With forty houses this meant that houses at Forgebank were able to be sold at market prices without any additional charge for the high environmental standards and the communal facilities.

Each family or person buys a house on a 999-year lease, and is free to sell it whenever they want. Any prospective buyer has to be voted in as a member of the co-housing group. Should the buyer not be admitted, Lancaster Co-housing can buy the house back after a fair valuation, and so be able to sell it on to others.

Lancaster Co-housing is the owner of the site and the buildings. There is also an old mill, which is being managed by a separate holding, called Green Elephant (GE). This is a management company that rents out space to people needing workshops, office or storage space. GE has been operating for less than a year, but looks to be breaking even already at this early stage.

LCH also have other community enterprises, including a food cooperative and a car pool. The food cooperative is completely informal, run by members volunteering their time to order wholefoods and make them available to other members in a locked storeroom to which all members have access. With an annual turnover of over £20,000, the food co-op will probably have to be formalised soon. Another enterprise is the car pool, which is legally registered, now owning four cars, including one electric car. Members pay according to both time and distance driven, and membership is open to people living outside Forgebank.

The terraced houses next to the River Lune under construction in 2012.
Photo: http://www.lancastercohousing.org.uk

In addition there is a shared laundry, with two washing machines operated by tokens available at the co-housing office. There are communal guest rooms and a bike workshop and storage space. Heating is by a district heating scheme, housed in the GE mill, and running off woodchips. In combination with highly insulated houses the heating costs are cut to an absolute minimum. This also means that the houses can be relatively small, not needing room for big heaters, washing machines or guest rooms.

The common house has a large communal kitchen where members take it in turns to cook evening meals. The community eats together three or four times a week, and the common house is used often for coffee and just hanging out. Opposite is a children's playroom, so the whole family can spend time being with others. In the spring of 2014 the Forgebank community was about sixty adults and fifteen children, all the houses were filled, but the mill was still half empty.

For more information: http://www.lancastercohousing.org.uk

6

Design is also celebration, this is from 2011.
Photo: http://www.lancastercohousing.org.uk

Process design in 2010, planning the terraced houses.

Tips for facilitators and groups

Brainstorming and wild design

Design is creativity in action. In order to stimulate the creativity needed for good design, I often use a brainstorming session, this is a very good tool to get the creative juices flowing. I ask participants to call out ideas and connections for things to be designed, often related to a theme that the group is focussing on. A certain degree of trust has to be established within the group, and a couple of ground rules laid down. The first is that there should be no criticism of any kind. The second is that each and every suggestion should be greeted with cries of enthusiasm and a round of applause.

Ask for two volunteers who can write clearly to note down the suggestions that are called out.

Make a list of between ten and twenty keywords.

Divide the participants into small groups of between two and three people in each. Each group chooses one of the keywords as a starting point for a design. I ask them to use the design tools and principles that have been introduced, and to use their imagination and creativity to present something. Each group then has ten to twenty minutes to work out a quick design.

In plenum, each group presents the design and then gives a quick explanation of how they arrived at the design. What design tools did they use, how was it? Give the group as a whole an opportunity to ask questions, clarify points and make comments. Feel free to add ideas, or quick explanations of similar designs.

The whole exercise takes about an hour, depending upon size of group.

This is a good tool to use before they start a design, to get their creativity going.

6

Outdoor dining area at
the only organic agriculture
school in Norway
built out of local timber.

Chapter 7. Building

Good building design means balancing available materials and ambient conditions. The previous chapter focussed on design criteria and the context of the building. This chapter looks at what kind of materials would be used in building, what kind of qualities each one has, and how they can be employed.

Materials

Globalisation has made itself at home in the building world, as in so many other spheres. What is suitable for Rangoon is also thought to be good in Reykjavik. Large buildings have become 'statements' by superstar architects who seem to be more interested in their status as artists than in the building's environment. In this development the context of the building has often been forgotten.

One way to overcome this ignorance of context is to look at local indigenous traditional techniques. What materials did they use, how did they use them and are these materials still available? Has the environment changed, have the forests been cut down, the quarries built over or used for landfills? By observing how people fulfilled their need for shelter in the past, we may be able to find more appropriate materials for use in the future.

Most houses built in the western world today rely on industrial materials: steels, alloys, plastics, various forms of artificial stone, and wood that has been shaped and impregnated, often chopped up and put together again. Mining, treating and shaping these materials requires different amounts of energy, and usually extra transport, all of which has an environmental impact. We call this the 'Embedded Energy' of the materials. If we take as our starting point the cost of felling, sawing and transporting timber, estimated at 580 kWh/tonne, we can compare other materials:

Timber	580
Bricks	2,320
Cement	2,900
Plastic	3,480
Glass	8,120
Steel	3,920
Aluminium	73,080

As you may see from this table, these are values that are in different dimensions. Comparing wood and steel, we may appreciate how much embedded energy is needed for a steel frame building as compared with a wood frame building. We will discuss these different materials in more detail.

Stone

Because of its weight, transporting stone over long distances is expensive, but if you have stones lying around on your site, you can save lots of money. However, stone requires skill to work with. Unless you are an experienced dry-stone wall builder you need to use a cement matrix to hold the stuff in place. You need to be aware of the geology; some stone is hard and difficult to work with, other stones are soft and erode easily. Slates split easily but can be used for roof tiles or paving slabs.

One easy and effective way of using rough stone is to build a box of wood and fill this with stones, cement and reinforcing. Put the nicest looking and largest stones against the wooden framing for a pleasing effect. Once all the stones are in place, just pour in the cement. These wooden boxes can be built in sections which are dismantled as the cement sets, and which can then be moved along to make the next section.

Rough stones require great skill to build with.

Easily shaped soft stone like this from Iceland can be cut to fit together.

Earth

Earth has been used in building for thousands of years in many different ways. One of the simplest is cob building, just placing handfuls of wet earth, a mix of clay, sand, silt and chopped straw, into flowing shapes that may be decided upon as you go along.

Adobe is earth bricks, dried in the sun, and cemented together using the same earth mixture as a bond. Pisé involves building a framed box, ramming damp earth in hard, letting this dry and then moving the box upwards to then ram in the next layer. In all these cases with an earth plaster finish on the inside and outside, you have a heat retaining wall, thermal mass, breathing qualities to give good air and a healthy inner climate.

You can make a test of your particular earth by taking a sample, shaking it in a jar of water, and letting it settle for a few hours. The coarser material will settle at the bottom, the finer in the middle, and the organic matter on the top. Ideally a mixture of between 50–70% sand, 10–20% clay, and a good addition of chopped straw is the best for an all-round plaster. The simplest plaster I've used was just a mixture of earth, sand and a little cow dung to make it hard.

In traditional Icelandic building, stone and earth are used as alternating layers.

Earth dome at a Permaculture project in the Northern Negev desert in Israel, for volunteers at the project.

Bricks and blocks, tiles and pottery

By including a much higher percentage of clay, and subjecting this to high temperatures, we turn the material into something that resembles stone; a fired brick. This requires much more energy, but the product can be as durable as stone, and bricks have been found which are thousands of years old. Ordinary modern bricks are standard sized, but you still need skill and experience to lay good walls. Fired roof or wall tiles are waterproof, and will last for hundreds of years.

By adding cement and using other materials instead of earth, you can skip the firing and use the chemical hardening process instead. These blocks can be made in many shapes and sizes, and make the building process much easier. They are typically much larger, and may be shaped for things like corners, strengthening courses and door or window surrounds.

Making your own cement blocks is also possible. In the 1950s Raul Ramirez of the Inter-American Housing Centre (Spanish acronym: CINVA) in Bogotá, Colombia, developed a lever-operated press to make bricks from earth. The Cinva Ram, as it was called, has more than half a century of experience behind it and tens of thousands of homes have been built with it.

Similar earth or sand blocks can be made with a small percentage of cement added, requiring less compaction. Tiles for roofing can be made in the same way.

Somewhere I once read about an experiment where a pile of dry firewood was covered in clay, with a few openings, and the wood was set on fire, thus firing the clay into a pottery dome, a kind of igloo. Afterwards you just rake out the ashes and move in. Try it and see what happens!

Fired and glazed tiles have been in use for hundreds of years, and Arab architecture has developed styles that are truly amazing in complexity and beauty. Many people use glazed tiles in kitchens and bathrooms. Great mosaics can be made using broken pottery in a cement matrix.

Glass

Some modern windows are very high-tech, very expensive, and really minimise heat loss. At that end of the range, with aluminium alloy or plastic coated wooden frames, triple plated, vacuum spaced, possibly photosensitive glass, you need to consider the embedded energy and the processes that go into the components. However, windows typically account for an enormous amount of heat loss in housing, not just the glass but the fittings and tightness of the frames. Here in eastern Norway, with normal winter temperatures down to minus 20 Centigrade for weeks, sometimes months on end, it may well be worth considering hi-tech windows to reduce energy loss.

At the other end of the range, scavenging windows and frames from houses about to be demolished, you can find the cheapest and most environmentally friendly solution. Get hold of the windows before building the house, so that you can plan the window sizes right from the start. This is highly recommended in more temperate climates, another good example of how important it is to consider the context of the building rather than to lay down dogmatic rules.

Metals

Today we use a wide range of metals in our houses, both for structural purposes, and fittings and connections. They are precious resources, mined, heavily processed and often transported great distances, with all the embedded energy problems that this entails. On the other hand, with careful forethought they can often be recycled quite effectively, and that has an enormous bearing upon the energy use. Anything you can rescue from a landfill will contribute to environmental improvement by reducing waste and the extraction of new materials.

It is worth bearing in mind that certain metals represent serious health hazards. Lead in pipes and in paint is perhaps most well known, but there are many others: mercury, nickel, zinc, silver, arsenic, and so on.

Timber

Timber fresh from the forest is clean, can be shaped with fairly simple hand or power tools, can be fitted together with joints or pegs, usually ages well, and can be recycled in many ways when we have finished with the building. Using timber is a skill, and some types need to be dried for a time before use. Other types of construction call for freshly felled timber that will dry into the required shape. This is especially true of log construction, where the weight of the roof will press the logs tightly together as they dry out.

Building with whole logs in Norway. The building on the right is over 400 years old: the one on the left was built about twenty years ago.

Traditional Norwegian log building.

Various kinds of timber have different qualities, some withstand water by their very nature, some split easily giving us shingles for the roof, others are soft and easily worked into decorative elements. They come in many colours and often with delightful graining. Combinations of different timbers can be a real joy to live amongst.

Try to avoid impregnated timber. Many modern builders seem indifferent to this, not seeing the paradox of having to dispose of the waste as 'dangerous materials' while at the same time it seems fine to them to live with such treated timber in the house. Timber treated with arsenic to avoid fungus rot is highly toxic, and can't be burnt to heat the house, whereas natural timber waste is fine for heating. One way to preserve timber going into the earth is by charring. This is an old tradition, and charred timber will succumb to rot much more slowly than uncharred. Just let the ends of your fence posts lie in a good fire for a little while so the outer surface becomes black.

Grass and stuff

Canes, reeds, bamboo, grasses and straw all belong together. Straw houses are as old as the earliest shelter building. Examples of vernacular architecture from around the world include Mudhifs in Iraq made of reeds, Balinese bamboo houses and the reed shelters used around Lake Titicaca in Bolivia.

Straw bales (SB)

Strawbale buildings have become an iconic Permaculture technique, sometimes ridiculously so. They are only really sensible where straw is a plentiful waste material from cereal growing, and where there are old-fashioned balers that can turn out the traditional old manageable bales. One of the problems of agricultural development is that many farmers in the western world have gone over to big bales, which weigh anything from half to a whole ton, and cannot be moved by hand.

The biggest threat to good SB construction is from water and humidity,

Straw bale house in the first, temporary, building project at Hurdal Ecovillage in Norway.

not fire. Experiments in various countries have shown that the tightly packed straw may smoulder for a long time, but will not flare up in an open fire. In fact, these bales are far less of a fire hazard than a standard timber frame building.

SB buildings do need good footings to raise them above any possible rising damp from the ground, and good eaves around the roof to keep the rain off. In addition, good plastering both inside and out is absolutely necessary, with careful attention at corners, tops and bottoms so the bales are completely sealed from possible intrusion of rodents or other destructive life forms. In one building we made here in Norway, cows came into our building site and grazed a corner of our wall before we could get it properly plastered!

Modern SB houses have been standing for nearly a century already. The original load-bearing construction of the Nebraska SB house is being increasingly replaced by a post and beam structure where the bales are used as infill. This has two distinct advantages; the frame can easily fulfil

7

building code requirements, and the frame and roof can be built first, so you can move the bales under the roof and carry on working even if rain threatens.

SB houses have the advantage of being built from an organic, renewable material. They breathe, depending upon the plaster you apply, and have an insulation value far higher than generally demanded by building codes. The actual wall construction, both bales and plaster, requires a group effort by people who need no special skills, though they do need an experienced guide. This makes SB construction really appropriate for an Ecovillage, where members can help each other build their homes, and the building process can be turned into a celebration of community.

Natural fibres

Today the tipi and the yurt have made a comeback in alternative circles, and are viable forms of quick and relatively cheap housing. Wool, cotton, linen, silk, sisal, jute, rayon, kapok and coir are all worth considering. One big advantage is that they generally by-pass building codes, not being buildings, having no real foundations, and are essentially temporary structures. For an Ecovillage that is developing a seasonal course and seminar centre, buying or building half a dozen such structures would be a small investment, soon paid back by participants' fees.

Hemp used to be produced in vast quantities in many parts of the world and was the basis for rope and coarse material for literally millennia. As a crop, it requires very little protection against pests and diseases, and as a product, it would fit perfectly into a system that wants to grow and use organic, natural fibres. A good example of a wholly contained and value-adding industry for an Ecovillage might be to grow hemp, turn the fibres into cloth, and sew tipis and yurts for sale.

Generally, fibres would form components of your indoor fittings rather than the materials from which you build, for example: curtains, carpets, wall hangings and lots more, not to mention clothes, bedclothes, towels, and so on.

Tipi, traditional North American housing solution, portable and light.

Paints

Paints consist of a pigment floating in a liquid, either water or oil, and today can contain many compounds that are a hazard to our health. There seems to be an increasing awareness of these hazards, and some paints are now marked as suitable for allergies, others with an environmentally friendly logo. There are also small businesses specialising in totally toxic-free paints.

If you can, find these suppliers of natural paints and varnishes. They exist in small numbers, and are following an expanding pattern not so different from the organic food market of a couple of decades ago.

You can also mix your own paints, the accepted way is to buy cooked linseed oil and add pigments. A paint that I have used a lot is just linseed oil, water, oatmeal to emulsify the oil, and iron oxides for the traditional red Swedish country barn colour. In Scandinavia this can be bought direct from the makers, Falu. This is a breathing paint, and will help keep your indoor ventilation fresh.

7

Junk, recycling and earthships

All over the world we are knocking down old buildings as fast as possible, and often many usable components are shovelled into landfills. Planking, beams, windows, doors, plumbing fixtures and furniture can all be salvaged and reused. This is not appropriate for the mass-produced modern industrial house, but rather to the individual, tailor-made, hand-built house. This can be ideal for an Ecovillage, creating a cottage industry, collecting materials, storing them in an organised way and then using them. The ecological saving is significant and reduces the embedded energy costs of a building. For example, at Earthaven Ecovillage in North Carolina, one building was built using crates thrown out by a local business, the free materials determining the shape and construction of the building. Hurdal Ecovillage in Norway made a deal with a local manufacturer of doors and windows to buy discontinued models and designed many of their initial buildings around the sizes and shapes of these windows and doors.

Floor made out of broken bathroom marble that had been discarded.

Building roof trusses being constructed out of recycled timber.

One well known concept within the Permaculture and Ecovillage networks is the 'Earthship' developed by Mike Reynolds in New Mexico during the 1960s and 1970s. Old car tyres are stacked up and filled with earth, creating massive thermal walls. They are then plastered outside and in, using earth as described above. Tin cans are cemented together with mud to make insulated interior walls. The idea of using junk not only makes our buildings cheaper, it also fills a need in getting rid of and making use of an energy-embedded resource. In Permaculture there is no such thing as waste, only resources looking for a use.

Earthships are more than just junk. They are designed to be self-sufficient in water, catching the rain off the roof; in energy, generating heat and electricity from the sun; in food, by growing edible plants in a greenhouse along the south-facing wall (in the northern hemisphere). They also take care of their own waste by treating the waste water to extract the nutrients, and recycling as much as is possible of consumer stuff.

Plastics

Most plastics are made from oil, and once made can be recycled almost indefinitely. This makes it a useful and flexible material. This is a far better use of oil than just burning it to carry us around to places we often don't really want to go to anyway! The biggest drawback to many plastics is that they contain gases that leak out slowly. Another drawback is that nearly all plastics give off poisonous gases when burnt, and are heavy industrial products. Given that we can minimise these two problems, plastics have a place in our list of building materials.

If we want electricity, I still can't find a way of getting round the insulation sheath around the wiring needed to carry the electricity around the house, which is usually made of a plastic-based substance. Water tubing presents the same challenge, and there is no doubt that modern, inert, plastic pipes are better than lead ones.

A certain amount of commonsense pragmatism is required here.

A checklist of materials for sustainable building

There are no defined laws of do's and don't's when thinking about materials. Everything has to be weighed by balancing the embedded energy involved in manufacturing and transport against the savings in energy use in the building over time. So high-tech windows may still come out on top by the energy savings over several decades.

It might be worth listing some general considerations:

- Avoid ozone-depleting materials
- Avoid materials with high energy costs
- Utilise local materials
- Look for locally recycled materials
- Minimise use of old growth lumber
- Minimise materials that discharge gas
- Minimise use of pressure treated wood
- Minimise packaging and resulting waste

Deeper considerations

The materials we use for our buildings have their unique qualities, not only in the material sense of stone being massive and solid, while cloth in a tent or yurt seems light, airy and transient. Materials also have spiritual and social qualities. Modern industrial components often require skilled assembly workers, but sometimes lack the feeling of a lovingly handmade item. Natural materials usually age better than synthetics.

These are qualities that we need to consider in our design and choice of materials. Will this material enhance our social and spiritual lives? Will it give us opportunity to celebrate community together, or do we have to bring in a group of professional outsiders to construct the building?

Imaginative building giving good placing for the solar panels.

Community profile

Friland

The Ecovillage at Friland was established in 2002 and won national fame through a project partnered with the Danish radio and TV consisting of building a meeting hall called The Raven. This generated a large amount of very positive publicity and people started moving in, building their own houses out of very cheap materials. The first regulation plan consisted of twelve building plots. There are now about fifty houses in the village, with more being planned and built. They have between seven and ten thousand visitors each year.

Friland lies in Northern Denmark, 35 kilometres east of the city of Aarhus. The local municipality was supportive, and made land available from the start. Friland has been actively involved in spreading both the Permaculture and Ecovillage design concepts in Northern Europe, hosting a number of courses and events: Nordic Permaculture meetings, festivals, and training courses.

There are a number of principles which members are asked to follow, which has made this Ecovillage quite different from many others. Each person or family is expected to build their own home without debts or loans. They are also expected to start an independent business, and receive no hand-outs. Each house cleans its own wastewater and everyone is expected to reduce their use of energy and resources.

Read more on: http://www.friland.org/?page_id=2374

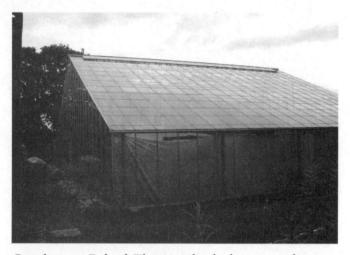

Greenhouse at Friland. The owner has built a more or less conventional house inside the greenhouse, giving him a much warmer climate.

Timber and straw bale, with a living roof for maximum insulation.

Tips for facilitators and groups

Solar tracking exercise

You will need string, a small box with a lid, balls/oranges/other small objects, and a group of between ten and fifty people.

Put the box representing a house in the middle of the space, then align north, south, east and west.

Ask them where the sun comes up in winter and mark the direction with the ball. Do the same where the sun sets in winter, and mark with the ball. Have one person stand between the two as a mark for how high the sun is in winter at midday. Now stretch the string from ball to ball via the midday sun, and arrange the group along the string to make a bow form. Ask them to eyeball the angles of the sun on the box, and indicate overhangs, east and west windows and any other features they might like to have on the house.

Now repeat the exercise with the summer sun.

7

You can now check the architectural details of windows, overhang, sitting areas, sleeping room, storage, kitchen, living room. Get the group to add in views, wind direction, access, for more details.

Greenhouses at Solheimar in Iceland heated with geothermal hot water yield plenty of good vegetables in an otherwise challenging climate.

Chapter 8. Energy and Technology

Our requirements for energy, and the kinds of technology we use, are major factors in our impact upon the environment. We might choose to do without modern technology, like the Amish in the USA or Tinker's Bubble Ecovillage in the UK. Or we could decide to use the latest and the best of modern technology. We have choices, and the important thing is to be able to make decisions from a position of knowing how different kinds of technology and energy sources affect the environment.

Critique of nuclear with emphasis on Fukushima

Nuclear energy has been one of the most disastrous experiments of modern times, and it continues to amaze me that some people still view it favourably after the Chernobyl and Fukushima accidents. Safety apart, just the economics of nuclear energy is shrouded in lies and deceits. A note in *The Guardian Weekly* in April 2011 reported that governments around the world were pledging $US785 million to help pay for a reactor shield around the Chernobyl plant. This brought the total raised for repairs at Chernobyl to $US1.8 billion. This is a direct subsidy to the nuclear industry, had it been added to the cost of nuclear energy, it would have made it much more expensive. Instead of paying for what it really is costing us, nuclear energy is subsidised to the hilt, making its price completely false.

In the same issue of *The Guardian*, it was further reported that after 25 years, the long term health effects of the accident at Chernobyl had still not been fully researched, and called for research to follow the health of

600,000 children living in Belarus, Russia and Ukraine. Keith Baverstock, a former World Health Organisation official, estimated that this research would cost a further $US14.5 million.

Two months later the same newspaper reported that the Japanese cabinet had decided to transfer the country's Nuclear Safety Agency from the Trade Ministry, dedicated to the expansion of the nuclear industry, to the Environment Agency. This was prompted by revelations of lies, misinformation and 'dirty tricks' carried out long before the Fukushima accident, and in the immediate aftermath.

Not only is nuclear energy dangerous and costly, but the people running the industry are not to be trusted to tell the truth about it by warning citizens of direct dangers that threaten them.

In a situation where professional economists refuse to listen to the world as it is, and politicians don't exercise their responsibility to inform citizens of future dangers, we need to take responsibility ourselves, something that is a defining principle in Permaculture. Even though I like to think of Permaculture as pragmatic and not dogmatic, it is my firm and irrevocable assumption that Ecovillages would not consider using or contributing to nuclear energy in any way.

Solar power

Solar power already does most of our heating. Absolute zero, at minus 273°C, occurs in outer space where the sun is too far away to be noticeable. Most of our planet is already up to an ambient and relatively comfortable temperature in the zero to 30°C range. We regard minus 40°C and plus 50°C as outrageously extreme. What we need to do is to adjust the final temperature a few degrees up or down.

Solar energy is coming in as heat, so it makes most sense to use it to heat our buildings and water.

The simplest technique is coiling a black water hose on the roof on its way to the shower and you'll have hot water for washing in whenever the sun shines. Next step up in sophistication might be to make the collector a little more efficient by enclosing it in a glass fronted, insulated box with a black background and bringing the heated water up into an insulated tank for storage. Warm water rises up into the tank, and stays there till you need it. Install a pump, expand the size of the collector, and you might circulate the hot water round your building through some radiators or even under the floor. Make sure that you don't get undesirable side effects like the whole thing working in reverse at night, when the collector will cool down below the temperature of the house. Either you will have to turn the pump on and off yourself, or install some thermostats to do it for you.

The easiest way to make electricity from the sun is a photovoltaic (PV)

Solar energy, both for heat and for electricity, is going to take over the energy market, and the big companies are well aware of this.

collector. Just buy it, set it up facing the sun and plug it in to converters, batteries or the grid. Electricity is easily transported, can be utilised in a wide variety of ways to do work, and be stored in batteries. However there are some problems with PV collectors. Their production requires a lot of energy and has some dubious pollutants. They are relatively expensive, but on the other hand seem to last a long time, and the price is coming down all the time. Solar energy installations are growing fast. Globally around 37 GW (gigawatts) were installed in 2013, a rise from 30 GW in 2012, and expected installations in 2014 are going to deliver an added 45 to 55 GW.

Falling water

The power of falling water has been harnessed for many centuries, powering corn mills and sawmills. In more recent times it has been used to create hydro-electricity. In order to have a supply of water, you need to dam up rivers, creating lakes that can be piped down to the actual turbines. Lakes are created in wilderness areas, rivers are diverted, sometimes through mountains and across watersheds. Some valleys lose their rivers, other valleys are lost to lakes. Today it's hard to find a river or lake in Norway or Sweden that hasn't been dammed or regulated. This is not so benign as it is sometimes made out to be. Very large dams in some regions take out farmland, displace thousands of people and contribute to damaging climate change.

Another approach is to create microsystems. This means building very small power plants, and using water from relatively small streams without altering the natural drainage more than just putting a falling stream into a pipe to create a strong jet at the bottom. The contribution of micro power stations is enormous, estimated in Norway to be 10% of the total needs.

The amount of energy you can get from a small water source depends

Falling water has been used for centuries.

upon the combination of two factors: the amount of water running through the system, and how far it falls. Basically you get the same energy output from 100 litres per minute falling ten metres as from 10 litres per minute falling a hundred metres. From the point of view of which kind of machinery you need, the less water the better, because larger amounts require large machines. From this rule of thumb, it might be clear that the less water you have falling further makes for smaller, lighter hardware. So this is best suited for mountainous places, where lots of little streams flow fast down steep slopes.

Water power can effectively be used for mechanical power, and here it can be as local and as small scale as you want. For obvious reasons, location is the key concept. Mechanical power doesn't really transport over any distance. A mill for grinding corn or running a small sawmill has to be built where the power is.

Wind generators

Wind is the result of differential temperatures, the warming of the sun, and the cooling in the shade. Day and night, winter and summer, north and south, wind is the attempt to restore equilibrium after the effects of heat and cold.

The main trouble with wind power is that it seldom blows at a completely steady rate. If the power can be stored and then used as required, you have a really excellent source of reusable energy. The other option is to use it when there is wind, that's how windmills ground corn in the old days. When it was calm, the miller took a day off! Another way to get round this uneven supply is to use it solely for heating, which has a number of advantages. Just run the generated electricity straight into heating cables in an insulated water tank. Heat can be stored until needed, then pumped round your house for heating.

Of course wind is now increasingly used for generating electricity, and using the grid as a battery.

If you decide to go for wind the first thing you need to do is to check your site. All you need is to put up a recording anemometer, and read off the results. Either there is wind or there isn't. It will vary according to time

8

Clean energy from the wind.

of day and season, and you need to build up a picture. Again, you should do this for a year, because things can vary a lot. Weather measurements have been recorded properly for many decades now, and you can get in touch with the national meteorological service. Don't rely completely on these figures unless the wind recorder is actually on your site. Wind varies a lot with topography. A hill, a wood or any other obstruction can completely change the results. If you can take wind measurements on your actual site, do it for a few weeks and compare it with recordings taken by the national service. You might be able to see a correlation, and can extrapolate from that. You will also want to know from which direction

Wind is one of the cleanest most renewable sources of energy, but capturing it has a large impact upon nature.

winds are blowing, and this is usually displayed as a wind rose, showing graphically how much wind from which point of the compass.

Armed with these facts, you are in a position to consider your options. They end up very similar to the things we thought about connected with water power. Do you want electricity or mechanical power? If electric, are you going to store it yourself or hook into the grid. One option with wind is to use the excess to pump water into a storage dam. Then you have a considerable steady supply, given that your dam is large, and you can draw the water off into a turbine that gives steady power all the time. The more complex your systems, the more sustainability you create.

More than eighty countries now have installed wind energy systems, with a combined capacity of 282 GW (gigawatts) This is expected to double every three to four years and reach 1000 GW by 2020.

8

A total system

Combine solar, water and wind, and you have a really great system with some good backups. It would be a complex system, needing good engineering and sound maintenance, but at an Ecovillage level it might supply all your needs, with storage being the water in a high level dam, and the water turbine giving regular even electricity day and night, winter and summer.

In several countries you can hook into the national grid, and supply power from your own system, using the grid as a battery. This is probably the best option, and the way of the future. Once you step up to an Ecovillage scale things begin to look a lot different. First of all the scale changes considerably. Even a cluster of ten homes (which is a pretty small Ecovillage!), requires ten times the amount of whatever you need.

There are so many renewable sources of energy that we really shouldn't need to use fossil fuels, or at least only sparingly, or as an equaliser. From a short article in the *New Scientist* in June 2003, Fred Pearce quotes Karl Yeager, president of the US Electric Power Research Institute: 'Power grids will become more like the Internet – networks for sharing electricity among millions of independent domestic and community generators.'

Geological heat

Here we have two sources: hot springs and the earth itself. Unless you live in Iceland, Japan, New Zealand or other places where there are hot springs, forget about tapping into the first source. If you do, and have natural hot water bubbling up from down below, you're really lucky, and all you need to do is build a set of radiators connected with water pipes and send the water round the house, boosting with a small circulation pump if necessary. If the water is really hot and is coming out as steam, you are doubly blessed. Put a turbine on it, run a generator, let the steam cool down to water, then send that round your house, greenhouse or village, and let the generator power the pump.

Even if you don't have hot water laid on by geology, the ground itself can be a source of heat. In Scandinavia, even when the outside temperature drops to minus 30 Centigrade or below, the ground will only freeze to a metre or two deep at most. Below this the ground is then 30 degrees warmer than the air. So bore a deep hole, put in a heat extractor and concentrator, and you have a source of heat for your building or your water. As long as the system is modest, you have a free source of heat that is sustainable, non polluting and inexhaustible. If you go overboard and install massive systems, beware! You could end up with your house sitting on a super cooled block of ice! These systems have relatively high capital investment costs, but are really cheap to run.

Bio energy

Coal, oil and natural gas are the mineralised products of ancient forests and swamps. It took geology a long time and many complicated processes to produce these materials. We can bypass this process by growing the stuff ourselves. We won't get exactly the same complex molecular structures, but we can get oil, fuel and energy from various crops. Trees are the most obvious.

Wood

The best way to get wood is just to collect it from the woods, fallen trees, thinnings and dead branches. If you have old waste timber from buildings that need renovation or just are falling down, that's also fine. You need to check for paint or impregnations, though. The stuff they use for this is mostly a poison based on arsenic, so the timber should be treated as dangerous waste and not burned.

Coppicing is the traditional way to set up a sustainable supply of timber. It's tried and tested, been going in England for at least a thousand years. The idea is that the tree will establish a root system, and all you need to do is to cut off the tops, and the stock will shoot out new growth very

Burning wood releases captured and stored solar energy.

Firewood is a serious business in a cold country.

quickly. If you allow ten years for regrowth, you need to take out just a tenth of your coppice every year. Establish a good mix of trees, deciduous are the best (conifers don't regrow from their roots), and plan to take out a selected few from various parts of the coppice. That way you keep the character of the wood intact. On a small scale, this is best done by hand, with a horse, reducing the environmental impact even more. This could be a suitable industry at village level, reducing outside costs for complex and expensive machinery.

In recent years the burning of wood chips in larger furnaces for district heating has become more popular. The advantages are that it can use the whole of the tree, including branches and twigs. The furnace can be made highly efficient and heat is circulated in insulated pipes. It is an expensive system to install, but has very low running costs.

Fuel values of wood

Harder and heavier woods give more heat than lighter timber per cubic metre. Below is a list of the most usual kinds of fuel woods with kW values per loose cubic metre of material.

Rowan	1,606	Pine	1,177
Beech	1,528	Black alder	1,177
Oak	1,469	Willow	1,151
Ash	1,469	Asp	1,014
Maple	1,417	Spruce	1,014
Birch	1,339	Grey alder	962

8

Manure

Methane from manures and sewage is a good way to get the most from this material. There seem to be various points of view on this. In India the Go Bar Gas system became very widespread, a real alternative technology, using old oil barrels for the digester and old tractor inner tubes for the gas containers, with bits of piping connecting the thing together. In China similar digesters were constructed out of concrete, based on the same idea. Small scale and relatively easy to construct and run, they supplied gas for cooking or heating on a domestic level. They converted animal and human manure into a slurry that could be used on the land, and took out the methane which in any case would be given off in a less controlled form.

Biogas

Instead of converting vegetable residues (bio-mass) into methane in a digester, the material can be heated in a controlled atmosphere and will produce a gas not unlike propane. This was used in some countries during the Second World War to run cars, it was pretty inefficient but worked nonetheless, and with today's technology, could be improved on a lot. Conversion ratios of 70% to 80% are possible, and very little modification needs to be done to an ordinary petrol engine.

Vegetable oil seeds can be grown, pressed and used as fuel. It does not differ much from diesel in terms of burning characteristics, and it's not a big technical adjustment to switch your motor from one to the other. Growing crops for oil such as oil seed rape or sunflowers is only possible in certain climates. Up here in Norway I don't hold out much hope. Adding value to your own crops is always a good idea, and the waste from pressing oil can be used as animal food, or even fuel. Either way it can find its way back into the soil, as manure or as ash, or if neither of those two seem to be appropriate, just compost the stuff direct.

Alcohol can also be grown. Any sugar-heavy crop can be used. Sugar beets are probably the most efficient, and grow in cool climates, but there are plenty of other crops that yield sugar for fermentation. Barley springs to mind as the classic for beer and whisky, though it does require quite carefully controlled sprouting and drying. Whatever you grow needs to be put through a process of fermentation and distillation. Again you will end up with waste products that are organic and can be used as animal feed, fuel or direct compost. In the case of alcohol production, there will be other considerations of cleanliness and sterility that might involve chemicals or processes that could be problematical. You can sterilise with heat, steam for example, which is nice and clean, but demands quite high energy. Or you can use chemicals, which might be a problem disposing of afterwards.

In both of these cases, oil or alcohol, you will soon realise how much work is required to get the internal combustion engine ticking over and delivering work for you. This realisation is perhaps the most valuable result of all this. It's so easy with cheap oil: just fill up your tank and drive away. You don't actually pay for the long geological process of forming mineral oil underground from the swamps of the carboniferous period, and you don't pay for the pollution of the oil spills and the refineries, or the wars over oil resources. Growing and processing your own fuel will hopefully put the price per kilometre travelled up to a high and realistic level.

However, don't think that I am dismissing the idea of these home grown vegetable fuels. They can be stored and used as required, a really important feature. You might want to use a generator to power washing machines or power tools for occasional use. This would be ideal.

8

A word of warning about Biofuels

In its recent review of the Fuel Quality Directive, the EU proposed a default value of 107g CO_2 equivalent per megajoule of fuel (CO_2/mj) for oil from tar sands, as compared to 87.5g CO_2/mj for crude oil, reflecting the greater environmental harm that its production causes.

While advanced 'second generation' biofuels comfortably outperform fossil fuels in the EU's new data, palm oil is ascribed a value of 105g, soybean 103g, rapeseed 95g, and sunflower 86g, once indirect land use change (ILUC) is factored in.

The data propose ILUC-incorporating CO_2/mj values for biofuels as follows:

Oil from tar sands	107g
Palm Oil	105g
Soybean	103g
Rapeseed	95g
Crude oil	87g
Sunflower	86g
Palm Oil with methane capture	83g
Wheat (process fuel not specified)	64g
Corn (Maize)	43g
Sugar Cane	36g
Sugar Beet	34g
2G Ethanol (land-using)	32g
2G Biodiesel (land-using)	21g
2G Ethanol (non-land using)	9g
2G Biodiesel (non-land using)	9g

Unconventional energy sources

The following unconventional sources are listed here as starting points for discussions designed to stimulate creativity. They can be used as exercises.

Human body heat is generally not included in energy source lists. But in fact it can contribute seriously to domestic heating. In Sweden, not known for its hot weather in winter, there are superinsulated houses that have no other heating than human body heat, plus a little help from light bulbs, of the old fashioned kind.

Animal heat. Animals on the ground floor, humans above. An elegant design, but you might want to consider the smell factor.

Hay Box Cooking. This is also known as Retained Heat Cooking. A much more accurate term, as you don't really need a box full of hay to practise this. A saucepan of food, soup for instance, is brought to the boil, then placed in an insulated box and left to cool itself slowly over several hours.

Trees shaking in the wind. Include the creativity factor. This is a really good design exercise.

Bicycle power. Pedal power has been claimed to be the most efficient way of generating energy yet invented. Cycling is really becoming popular now, and more exciting designs include power generation, turning small machines like domestic corn mills, and I have even seen a design for ploughing, using stationary bicycles.

Doors. I had a participant on one of my courses who did a little design exercise on how he could capture the energy created by opening and closing doors manually. Everyone agreed that it might not be the answer to the world's energy problems, but the creativity involved in looking at wild ideas certainly proved valuable.

Carbon capture and storage

During the last couple of centuries we have overused our ancient store of fossil fuels, creating a radical change in the composition of the atmosphere and triggering climate change that is seriously challenging for our survival.

There are several high-tech solutions being considered, including installing mirrors in outer space to deflect incoming sunshine, or seeding for increased cloud cover. There have been continuing talks about carbon taxes and quotas. None of these are delivering an improvement. Back to Permaculture design, we find that nature has designed a carbon scrubber that works brilliantly with relatively little input from our side except to get it started.

Forestry, either in the form of planting or replanting, will scrub CO_2 out of the atmosphere, absorbing it and dividing it up into Oxygen which is reabsorbed by the atmosphere and Carbon which is stored in the lignins and cellulose which can stay there for decades and sometimes centuries, depending upon the tree and how it is managed. Massive tree planting

Planting and maintaining forests helps to capture carbon and offset climate change impact.

schemes have shown to be possible, in the USA in the 1930s, and today in Africa and in China.

Albert Bates has run the maths on re-afforestation at the The Farm Ecovillage in Tennessee and at Earthaven Ecovillage in North Carolina in his book *The Biochar Solution* and it turns out that they are both carbon negative, capturing and storing more carbon in their forests than they are using in their lifestyles. They are in effect showing us the way to the future.

I came across the following story about a town in Austria. In the 1980s National Economic Studies revealed that Gussing had become one of the poorest districts in the country. It was concluded that if the district could create its own energy from locally sourced, renewable fuels, it could begin to rebuild and strengthen the local economy, create jobs and bring more wealth and prosperity to the region.

It was decided that energy from trees was one of the most abundant sources of locally-available fuel so the earliest measures at the start of the 1990s included energy efficiency incentives and heat from biomass.

Since then a number of renewable energy technologies have been applied, including district heating from woodchip, solar thermal energy, wood gasification and the creation of biogas from farm waste.

These measures resulted in Gussing almost halving its CO_2 emissions since 1996 (even though energy use has gone up), the creation of over fifty companies, 1,100 new jobs and a total energy sale of 13 million Euros per year (2005) which now stays in the local economy. 44,000 tons of local wood are consumed each year, small farmers and foresters have a market for their lesser quality timber and the forests are maintained and expanding. (See *Permaculture Magazine* 77, autumn 2013.)

Bioregions

It may be possible to be self sufficient to a certain degree as a family, but it would be easier to achieve a high degree of self-sufficiency as an Ecovillage. How far this is desirable is another question. Expanding to a whole region may be more realistic and has a higher community value.

We might define a bioregion as being an area that has some common features when it comes to landscape, climate, vegetation, culture and resources. Bioregional thinking has been around for decades, and attempts are made from time to time to redraw boundaries, replacing the larger nation states by smaller, more autonomous bioregions. In our Permaculture thinking we may agree that a top-down political type campaign is less relevant than building up bioregional awareness bit by bit.

We could act locally to create an association of residents of a natural and definable bioregion. This could have a population that may vary between one hundred and several thousand. One of the best ways to do this might be to create a network of organisations, acting locally to create an association of residents of a natural and definable region.

A bioregion needs an ethic, based on culture, financial and physical resources. One idea that has been around for a while is that we aim to 'live within our watershed', with the awareness that exports and imports impoverish. We might look at the difference between sustainable and self-sufficient, and we would certainly want our bioregion to preserve its natural character and have a high degree of stability, while improving regional sustainability.

In this context, self-sufficiency is not desirable in itself. Looking at our collective needs within the region, and checking how many of these can be met by the resources within the region may give us considerable savings in transport, at the same time creating employment locally. Security and stability are qualities that are worth aiming for. Food security is becoming a well-studied concept, and in the globalisation that is occurring in our farming and food distribution systems, we are in danger of losing this security.

Working together with existing organisations is crucial. Rather than alienating them, they can be regarded as partners, and the project can be presented as a sensible way to solve existing and potential problems rather than a radical reorganisation of society.

Permaculture gives us many tools for organising locally, most of which are already tried and tested by the Transition Town network, and of course play a big part in Ecovillage planning, albeit on a smaller scale. Energy and technology are key issues here, as is food production and distribution that we looked at earlier. We shall also see how economics can make a large contribution to this type of bioregional development.

Using natural materials, and with a living roof to maximise diversity and insulation.

Community profile

Solheimar

Solheimar in Iceland presents itself as a self-sufficient community inspired by Rudolf Steiner and Permaculture, and is a member of the Global Ecovillage Network. It has been described as the oldest modern Ecovillage in the world.

The village was established by Sesselja Sigmundsdottir in 1930, after she had visited Switzerland and had been inspired by the work of Rudolf Steiner and Ita Wegman with children with special needs. At first they lived in tents around an old turf farmhouse. They used the abundant hot geothermal water on the site to cook and for heating, and took in poor and disabled children. After the Second World War the place became a home for disabled children and in the 1950s and 1960s became a home and village for adults as well as children.

The village lies in a shallow valley in the southern part of Iceland. Their water is a little below boiling, and even though they have tried drilling in several places around, there is no cold water to be had. All their heating needs are met by having the water run through miles of piping, from house to house, including greenhouses, workshops, guesthouses and other public buildings. This means that they are able to grow a large variety of vegetables nearly all year round. Solheimar pioneered organic and biodynamic farming in Iceland.

The village is not formally a part of the Camphill network, but has extensive informal ties with both Camphill and other anthroposophical institutions working with people who have special needs. Like them, Solheimar focuses on possibilities rather than limitations, and on reverse integration, with the society or fellowship built up around individual needs.

Everyone goes to work in a number of different workshops; bakery, gardens, shop and coffee house, guesthouse, weavery, woodwork shop and art workshop.

There is a new visitors' centre, fully environmental, described as one of the most sustainable buildings in Iceland, with full insulation, solar power and wastewater treatment. They run educational, sustainability and community programmes, and school visits in this building. The community comprises 43 people with special needs and the total population is something under one hundred.

Read more on: http://www.solheimar.is/index.php?msl=english

Raised bed permaculture garden.

Tips for facilitators and groups

Create a bioregion

This exercise works really well if there is a spread of participants from throughout a country or even internationally.

Divide people up geographically, trying to form groups that have either a regional common-point, or at least some similar geography. If there is no obvious regional spread, divide the group up arbitrarily, and ask each group to define a bioregion of their choice. Then give them at least twenty minutes, an hour is better, to come up with some broad solutions under the following headings:

- Food and food support systems
- Shelter and buildings
- Livelihood and support systems
- Finance
- Leisure
- Transport
- Social security

Bring everyone together, and go through, region by region, what kinds of strategies they have devised. This is a really good exercise to do later in the course, giving participants a chance to bring together a wide range of design techniques they have covered already.

8

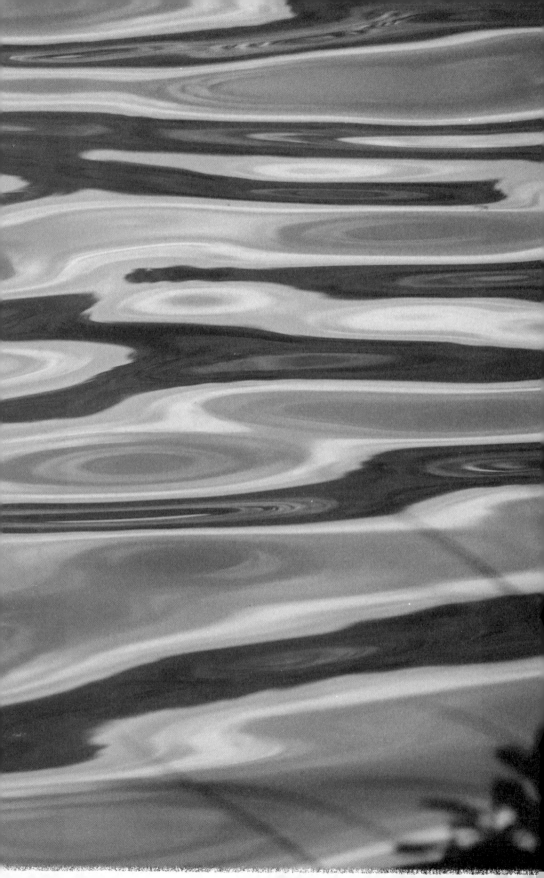

Chapter 9. Water

I think that all the designs I have seen that are based on the physical, material world involve designing with water. Kitchens, houses, gardens, farms and Ecovillages all need to address this element, and the patterns that water moves in can inspire other designs; of businesses, of educational programmes and of social themes.

Water on earth

The water cycle is an ecological pattern that many people are most familiar with. Water evaporates off the sea, forms clouds that travel inland and shed their moisture as rain and snow, which then collects in streams and rivers, flowing back into the sea. Thus water is in constant motion, circulating between the vast oceans and the fresh water on and in the ground. At any one time, only about 3% of this is fresh water, the rest, all 97% of it, is in the ocean, meaning that for our purposes it's out of our use, except to catch a few fish in and travel across its surface. This tiny 3% fresh water is vital to our existence. Our bodies contain about 60–70 % water, and this needs replenishing every day. As any of you who have been in the desert know, a few days without water means death.

All of this 3% water is not immediately available, however. About three quarters of it is locked up in ice sheets and glaciers, and a further 11% is deep underground in ancient aquifers. These are generally not being replenished, and so do not really constitute a renewable resource. They are usually very ancient, and being so long in the ground, are heavily mineralised, so when they are pumped up for use, constitute a long term health hazard if drunk, or contribute to salting of soils if used for irrigation.

The bit that's left is in groundwater, streams, rivers, lakes and ponds. That's what we need to live off. Clearly we need to take care of it, use it sparingly and keep it clean. We also need to share it between us, as lack of water will lead people to violence by any means in order to have access to it.

Looking after our precious water

We can store water in tanks as it comes off our roof during rain. We can also store it in the ground, by making sure there is penetration, and in ponds and dams. Plants, especially trees, have great water storage capacities, and in addition ameliorate the local microclimate, ironing out extremes of runoff, temperature and humidity. A forest is a lake above ground, its leaves slow down the impact of raindrops, and its roots hold the soil firm, restraining erosion.

In urban areas, runoff has to be carefully designed. On most roofs and paved areas, tarmac or concrete, there is no infiltration and any rain begins to run immediately, often accumulating to uncontrollable torrents within

Clean drinking water is absolutely necessary for life.

minutes or metres. Living roofs (turf roofs) can be of great importance here, slowing down runoff by storing the water in the turf and amongst the plant roots, letting it trickle out slowly for several hours after the actual rainfall. Garden beds along paths and roads have a similar effect but we might bear in mind that road runoff may contain pollutants from the tyres and engines of the passing cars.

Water reserves can be tracked by satellite. NASA is now collecting data, and a document from January 2014, reported in *The Guardian Weekly* a few weeks later, showed startling overuse of water reserves. So much so that California State Governor Jerry Brown announced a state of water emergency, and appealed to Californians to cut their water use by 20%. The same article stated that a billion people, one in seven of all people on the planet, do not have access to safe drinking water. Losses of water reserves are staggering, groundwater has been pumped out 70% faster during the last decade than during the previous one.

As designers, we have work to do!

9

The magic of water

The surface of our planet is about 65% water, something that is reflected in our own bodies. Water is a very special substance and there has been a great deal of research on this subject in the last few decades. It is the basis of life, it is lighter when solid (think what might happen if it was the opposite!), it flows easily and is always horizontal when still, and it's useful in solid, liquid and gas form.

The holy lake in Pushkar, Rajasthan in India, which attracts thousands of worshippers every year. Photo: Ruth Wilson.

Water has:

- Kinetic energy (when in motion).
- Potential energy (when flowing downhill with gravity).
- High capacity for heat absorption.
- High capacity for heat storage.
- Specific weight of 1 tonne per cubic metre.
- High acoustic transmission.
- Ease of spreading.

- High capacity to carry information.
- The ability to filter and mirror light.
- Simple molecular construction.
- Good therapeutic characteristics.

Flowing water creates highly dynamic patterns.

Uses of water

We might want to bear in mind the different ways we use water on a smaller, domestic or community scale. This will give us an idea of how important it is, and motivate us to take better care of our water.

We use it:

- For drinking, cooking our food, and cleaning.
- For transporting heat, warming our houses, buildings and greenhouses.
- For watering the garden and our household animals.
- To fill small ponds, creating microclimates and reflecting light into buildings.

- To create biological diversity by multiplying the edge effect between water and land.
- To transport waste out of our kitchens that we can use as nutrients in our garden.
- To fight fires.
- To create a cool store in our cellar by storing ice.

Washing in a lake in India.

A few Permaculture principles

Bearing all this in mind, we can formulate a few principles that we might take with us when we design for water:

- Use water as many times as possible as it passes through your system.
- Slow down the water flow across your property.
- Use gravity to move water around as much as you can.
- Solve the problem of contaminated water as close to the source of contamination as possible.
- Ensure that water leaving your property is clean.

Landscapes and keylines

With large site landscape design we can bring some of these principles into play. This is relevant for many Ecovillage designs, if the community is developing in a rural area. Observing water and how it behaves as it runs across the landscape is one of the most important design practices, and will give us information that we need when we design how we want it to behave.

Keyline analysis was developed by P.A. Yoemans and published in the early 1980s under the title *Water for Every Farm*. Bill Mollison bases a great deal of his landscape water design upon this pioneering work.

Keyline analysis is based on a detailed examination of the contours of the land to be developed. Familiarity with maps and map conventions such as contours is absolutely essential for any Permaculture designer. A contour line is an imaginary horizontal line linking together points that are the same height above sea level. As contour lines curve and swing, and as the distance between them increases and decreases, the form of the landscape will be revealed to the map reader. The point at which a slope changes from convex to concave is known as the 'keypoint', and 'keypoints' can be linked together by lines that may not follow the contour lines exactly.

Another important tool that can be brought into play here is the 'swale'. This is a ditch that is constructed exactly on the contour line. The opposite of a drainage ditch, which leads water away to a designated place further downhill, the swale catches the water and allows it to infiltrate the ground, building up the groundwater reserves.

Swales and drainage channels can be built along the 'keylines' allowing the water either to penetrate if the swales are horizontal, or leading the water where the 'keyline' dips, either into dams in the valleys, or to penetrate the usually drier ridges.

By constructing larger or smaller dams as the topography dictates along these 'keylines', runoff can be controlled and encouraged to penetrate into the soil, adding to groundwater storage. Using these principles, we can plan our site using the following advice where appropriate:

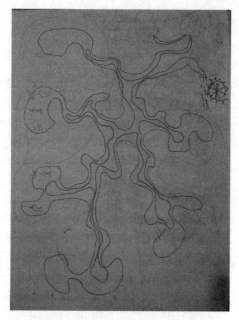

*Designing ponds to create maximum
edge between the water and the land.*

- It's better to have a few smaller dams rather than one big one.
- Dams should be situated in places that can catch water from as large an area as possible.
- Drainage ditches can be constructed to lead water to a dam.
- A smaller dam immediately upstream from a larger dam will collect silt and prevent the larger dam silting up.
- Dams can be multi-purpose, creating microclimates by reflecting the low winter sun.
- Dam walls can be used as road foundations, and wildlife areas.
- Damage from earthworking should be repaired as soon as possible.
- Topsoil can be removed and used elsewhere.
- Always think whole site planning, a dam is not just a dam, it's a part of the whole drainage and landscape picture.
- Construct windbreaks to reduce evaporation from dams and ponds.
- Site ponds and dams uphill from areas that you need to irrigate, then use gravity to move the water.

- You can seal a dam or pond with clay, if you have some on site.
- When irrigating, use drip methods; they are the most efficient.

Wastewater

Having collected our water, we use it in various ways, and afterwards need to clean it before it leaves our property. Biological wastewater treatment systems have developed greatly over the last couple of decades. We can separate our domestic waste into three areas:

- Solid human waste (humanure)
- Solid kitchen waste (food waste)
- Water from washing, together with urine (urine is sterile)
- In addition there will be solid waste such as paper, plastics, metal, glass. These should be sorted carefully and recycled.

These should all be treated separately, and in our Permaculture thinking, where there is no such thing as waste, only resources looking for a use, we will find ourselves in possession of valuable resources for our garden.

The solid human waste can be composted. This should be done with an extended compost cycle of at least two years. Then it can be used for nourishing the soil. Two years is generally accepted as being long enough for pathogens to break down naturally into their component organic elements. If you are concerned about composting humanure, you can apply it around fruit trees after the harvest, giving it many months to incorporate into the soil, and there will be no contact with the fruit whatsoever.

The solid kitchen waste can also be composted, and can be ready for use as a soil improver within a few months, depending upon the compost system you use.

Liquid wastewater from washing people or things, often called 'grey water', can be filtered into a wetland, growing a variety of reeds and other water plants. These can be harvested regularly and composted, thus returning the nutrients from the wastewater back to the soil. Urine

9

can be watered down and used as a soil stimulant. It contains high amounts of nitrogen, so watering down should be about 10% urine and 90% clean water.

A compost toilet can save thousands of litres of water over a year.

Flowforms

Flowforms were first produced by John Wilkes whose research on the characteristics possessed by water led him to consider its application in the treatment of sewage. Coming from a background in sculpture, Wilkes worked with Theodor Schwenk, the director of the Institute for Flow Sciences and the author of *Sensitive Chaos*. Schwenk had studied with the Cambridge mathematician George Adams, and all three had a background in Goethean science, developed further by their study of anthroposophy.

Water is a precious commodity. It is there to support life. Water is a mediator between the environment and the organism. In a meandering stream, a falling raindrop, a curling wave, a tumbling cascade or a swirling vortex, water exhibits a restlessness and potential to adopt a host of rhythmical forms. In all living things, rhythms are present. The pulse of blood through the arteries and veins, the alternating expansion and contraction of lungs, the peristaltic movements of an earthworm, the pulsating body of a swimming jellyfish all give expression to specific rhythms. In the body fluids of living organisms, order and rhythm predominate. As the main constituent of sap, blood, lymph and other bodily fluids, water circulates incessantly in distinct rhythms. Somewhere hidden in this phenomenon of rhythmical movement is the secret of water's quality.

Rhythm is generated by means of resistance, it assumes a 'breathing' regularity. The rhythmic meandering of a riverbed is formed by the resistance of surface and gradient to flow. The organs of the body are also created in a similar way. They then proceed to regulate the flow of blood and other bodily fluids, just as the river regulates the flow of water.

Flowforms generate rhythms in a comparable way by means of very specific proportions. The apertures or shapes through which the water flows are designed to resist the flow to such a degree that it hesitates. The flow momentarily loses its linearity or in other words its subjection to gravity. This unstable situation creates an oscillation that is then maintained by the shape of the Flowform.

Large flowforms are really great for kids to play in.

9

By building a variety of bowls across which the water would flow under gravity, it is encouraged to swirl in definite, lemniscatory (figure of eight) paths, based on patterns observed under natural, healthy conditions. Rhythms within and between the bowls can be attuned and harmonised to create a veritable symphony of movement. These were the bowls that came to be known as Flowforms. Preliminary investigations found that water passing across Flowforms became highly oxygenated and thus showed promise for wastewater treatment. The animated, rhythmical motions of water passing through Flowforms also produced a more ecologically vital condition. John Wilkes asked himself if such motions might enhance water's life-supporting capacity. If so, what combinations of rhythms would best meet the requirements of wastewater treatment? Would different ones be more appropriate for irrigation?

The use of Flowforms was first really applied in Camphill communities where wastewater treatment systems based on this research were installed in several countries in Western Europe, particularly the British Isles and Scandinavia. They show highly positive results, especially in the context of isolated rural communities.

The use of such systems in health clinics and hospitals has also been tried and found to be of tremendous benefit in the recovery of patients undergoing treatment. In the home, the office, the workplace and in enclosed public spaces, Flowforms have the capacity to humidify the atmosphere, and load it with health-giving negative ions, in addition to the aesthetic appeal of the sight and sound of falling water.

The treatment of sewage should be seen not just as a problem to be overcome; rather as a unique opportunity to demonstrate nature's remarkable qualities. Flowform cascades, ponds and created wetlands can be constructed with the aim of being quality amenities for the benefit of local residents. Rather than being placed out of sight and mind, they derive pleasure as well as an educationally rewarding experience.

Natural water treatment technology

Uwe Burke developed treatment systems at Oaklands Park Camphill Village in England in the 1980s. The system was designed to treat wastewater from a community of about one hundred people, incorporating a biodynamic farm and other food processing facilities servicing the village itself. The system was constructed with village labour, and began effective treatment in 1988, monitored by the local water authority. It was based on an existing septic tank with a capacity of about 12 cubic metres and a retention time of two to three days. To this was added a new settlement tank of about 5 cubic metres. After desludging in these tanks the sewage was moved to a series of open beds. The primary stage beds were planted with *phragmites* reeds. After that the water was moved to secondary beds planted with *scirpus lacustris, iris pseudacorus* as well as *phragmites,* and subsequently to the tertiary beds where *sparganum, acorus calamus* and *carexelata* were added. Aeration throughout the system was by using Flowforms.

The increasing complexity of the plant ecology reflects the gradual removal of various contaminants from the water. The final pond was fringed with a variety of plants, and contained carp, goldfish, bream and rudd. The total area used was about 100 square metres, about 1 square metre per person. The estimated cost of constructing the system was estimated to come to a per capita price of about £325, as compared to a mechanical (conventional) system cost of about £400. Most of the work was carried out by village residents, which reduced the real cost to about a quarter of that quoted. Since then aquatic sewage treatment incorporating Flowforms have been established in many if not most Camphill villages throughout the world, and have contributed significantly to the acceptance of these systems by local authorities.

9

Rootzone treatment

In the *Permaculture Drylands Journal* Michael Ogden describes a system he set up for a factory in New Mexico in the United States, a very different climate from that of temperate England. The wastewater to be treated was seen to be of two different kinds, the polluted water itself, and the thick liquid from the bottom of the septic tanks, called septage. This latter presented a significant challenge, composed of ammonia, grease, oil, hair and hydrogen sulphide gas.

The first treatment was in a pond, where aeration and mixing were done mechanically, and a greenhouse controlled the temperature and contained the smell. From this pond the wastewater was led outside to a reed bed planted with our old favourite, *phragmites communis*. Here the water was periodically pumped up, flooding the bed, and then allowed to trickle down through the gravel. This is a mechanical way to mimic the periodic flooding that takes place in natural river deltas, and it builds up a very specific set of micro-organisms. From this reed bed the wastewater made its way back into the greenhouse to a lagoon dominated by water hyacinths. Here most of the remaining pollutants were digested by the rich, largely anaerobic micro-organisms. The water hyacinths were harvested periodically and composted. Still in the greenhouse was a gravel bed, where the wastewater was further cleaned, to be led away outside afterwards into a further gravel bed before draining away into a meadow planted with native arid land species.

Roots extract waste materials from grey water, from Camphill Vallersund in Norway.

There are three principles being demonstrated here:

- More ecologies are better. Alternating pond, marsh and meadow gives a greater variety.
- Alternating aerobic and anaerobic environments is highly efficient.
- A breadth of varieties of species will treat the pollutants most effectively.

A word of warning

Wastewater may contain dangerous substances, not only pathogens from humans, but also toxic chemicals and medications. If you are a small family, in complete control of everything you put into the water, there should be no problem treating the waste appropriately. If you are a larger community with many visitors, it is another situation entirely. Visitors may have diseases, or they may bring dangerous medication, or soaps that contain detergents that will kill microorganisms. Workshops may use materials that leak heavy metals into the system.

If you are contemplating treating your own waste, it is absolutely recommended that you make contact with experts in the field, and find out what your local regulations permit. Many successful and safe systems have been put in place with a high degree of cooperation with local health authorities, and with a maintenance schedule that includes regular testing. The Permaculture Design course may give you some indications of which areas to explore, but can in no way be considered a recipe that teaches you everything you need to know in order to install such a system.

9

Vidaråsen celebrating its fortieth anniversary.

Community profile

Vidaråsen

Vidaråsen was the first Camphill community to be established in Norway, founded in 1966 by four idealistic individuals. Gradually they gained support from high school students who collected money for them by selling candles every Christmas. 'Lighting a Candle for Vidaråsen' became a byword for two decades, and collected enough money to build several houses and workshops through the years.

Vidaråsen has about 120 residents and lies a couple of hours' travel south of Oslo. The village comprises a farm, gardens, dairy, carpenter's, workshops in felt and painting, a weavery, bakery and a shop. They have a rich cultural life with concerts, theatre productions, lectures and meetings, and there is a therapy building for people who need extra care.

All their wastewater at Vidaråsen, both grey and black, is treated on site using ponds aerated by Flowforms. These three quite extensive ponds lie in a small valley below the village, and discharge into a nearby stream. The discharge is tested regularly and though not of drinking quality, contains no harmful substances.

Horses are still used for delivering things around the village .

Camphill Landsbystiftelse, (Norway's Camphill Village Trust) maintains six villages in Norway. They are all life-sharing communities, where people with special needs live together with co-workers. They are inspired by anthroposophy, the spiritual science developed by Rudolf Steiner. Art, handicrafts and culture are combined with a high degree of environmental awareness. Most of them have their wastewater treated biologically with ponds and Flowforms. At Solborg the ponds were tested over a period of a few years, to see how they functioned under the sub-zero temperatures experienced over many months here in Norway. They were found to be functioning just fine, the results were published

Flowforms help to clean the wastewater at Vidaråsen.

9

at an environmental conference in Sweden in 1991, under the title *Ecological Engineering for Wastewater Treatment,* edited by Carl Etnier and Bjørn Guterstam. This led to a change in the law in Norway, allowing small communities in rural areas to install similar wastewater treatment systems.

For more information: www.vidaraasen.no

Care home for the elderly.

Tips for facilitators and groups

An exercise in swale construction

This is a good exercise for a small group of between three and five people. If the group is much larger, it's best to divide up. It's useful to have some spades and shovels, and you will need some string, a weight (a rock will do), a marker pen and a number of short sticks. A knife is useful for cutting the sticks.

Cut three long sticks that are stout enough not to bend, and lash them with the string into an A shape. Stand the frame upright and attach the string to the top of the triangle, with enough length to hang below the crosspiece. Tie the rock or weight to the bottom of the string. This is then your levelling triangle, the Swale Marker. To calibrate, stand the triangle on some level ground, note exactly where the two legs are standing by pushing a small stick into the ground next to the leg. Let the string with the weight attached hang down steadily, and make a small mark where the string crosses the crosspiece. Now reverse the legs of the triangle and mark again where the string crosses the crosspiece. Now make a very clear mark exactly between the two marks you made, and whenever the string rests on that mark, the two legs will be exactly level.

You can now start marking out your swale, which, if you follow the path of the legs of your Swale Marker, will be exactly level. First determine where you want the swale to be, and start at one end. If you are two teams, start in the middle and work outwards. With two sticks, mark where the legs are by pushing them into the ground next to the leg. Now swing one leg around to where you think the swale should go, moving it back and forth until the string rests against the mark. Push one of your marker sticks into the ground next to the leg and swing the second leg along. In this way you can follow the contour along the slope, each marker stick will be exactly level from the previous one. The digging crew can follow on, digging a shallow ditch and throwing the spoil from the ditch into a little mound on the downside of the ditch.

You will soon have a good swale stretching across the slope of the hill, ready to catch any rain that will fall, and allowing it to infiltrate the ground.

9

An English village surrounded by its farmland. Natural capital.

Chapter 10. Alternative Economics

Economics can be defined as the system that produces and distributes the goods and services we need. We have never, in human history, had so many economists as we have now, we have never subjected economics to so much research and study, and yet we can't get it right. Many of these people are probably concerned, decent human beings, but the ethos of our practised economics is suffused with greed. Instead of relating to the words 'distribution of goods' economics today has to do with 'concentration of wealth'. Let's look at some statistics.

According to *The Guardian Weekly*, in 2013 the world had 1,426 US$ billionaires, a rise of 210 since 2012. The total assets of these 1,426 people were US$5.4 trillion, up from US$4.6 trillion the year before. The richest person was Carlos Slim, worth US$73 billion, followed by Bill Gates with US$67 billion.

In the 1970s the average American chief executive made about 42 times as much as the average worker. Today that figure has climbed to 380 times as much! (*Guardian Weekly*, November 2012.)

According to Oxfam in January 2014 the richest 85 individuals owned the same as the poorest half of all the planet's people, 3 billion of them. Forbes magazine updated this a few months later to 69 of the richest individuals owning the same as the poorest half. In the USA the richest 10% have captured over 90 % of the economic growth since the recession.

The myth of progress

During the last few hundred years we have developed a view of the world which sees the progress of civilisation as a linear progression from a savage and brutish Stone Age existence to an ever more civilised future. This is not a paradigm that is common to all humanity. Many cultures have looked back to a golden age in the past, and see history either in terms of waves, with golden ages coming and going, or as a gradual descent into chaos and evil.

In the late 1960s and early 1970s Marshal Sahlins, professor of anthropology at the University of Chicago, researched economics in societies that practised a Stone Age technology, hunting and gathering, largely nomadic, and with a very low material culture. He researched in both archaeology and in contemporary cultures that were similar. His book, *Stone Age Economics,* revealed a picture diametrically opposed to our prevailing paradigm. People in Stone Age cultures worked from two to five hours a day to satisfy their material needs of shelter, clothing, food and tools. They generally slept for a while during the day. They knew their environment and generally did not store food over long periods, trusting that nature would supply them with what they needed out of its own abundance. Of course there were conflicts, occasional natural disasters, droughts and floods that would lead to starvation and catastrophes, but these were generally occasional, isolated events. Comparing that world to the world of the 1960s, Sahlins estimated that 30–50% of modern humanity went to bed hungry every evening, and remarked that in our time, with all our technical, social and economic progress, we have made starvation an institution: '... hunger increases relatively and absolutely with the evolution of culture.'

Economic background

Economics is a social science that started with Adam Smith (1732–1790) who studied the connection between buyers and sellers, and discovered what is known as the 'invisible hand of the market place'. Since his time

Basic economic activity. A street trader in India.

economics has become an established study, and we can define three main developments.

Classical economics is the foundation of economic thinking, describing how the prices of goods are defined by the relationship between supply and demand. However, this type of early economic thinking did not create tools that national governments needed to intervene and reduce the impact of the busts and booms that nations and regions were experiencing. Neither did it address the fact that the real economy does not function like a machine. Players make decisions and react to the market in a variety of ways that are difficult to predict, and cumulatively produce results that often come as a surprise. Sometimes economics ends up in financial crises. It sees labour, raw materials and energy as just 'raw materials' and not living entities with rights and qualities of their own. For example, the scarcity of a raw material is not only a cost factor, but will affect future generations negatively. If we use up all the oil in the world, our grandchildren won't have any. Similarly, labour is not just a cost analysis

10

commodity, but real human beings with needs and desires. In a similar way little account was taken of pollution and its long term effects by these early economists.

John Maynard Keynes (1883–1946) developed a number of tools, embedded in solid economic theory, which national governments could use to ameliorate the negative effects of busts and booms. These were based on controlling national interest rates and using public spending to get through hard times. Keynesian economics built up the western world after the destruction of the Second World War, and was the basic foundation that created the affluence of Western Europe and North America up to the 1970s.

Monetarism replaced Keynesian economics in the 1970s. The US dollar was taken off the Gold Standard in 1971, and the planet was gradually encircled with trade agreements between rich nations, the World Bank, and other global institutions. Monetarism is associated with a right-wing type of thinking that considers survival of the fittest as best suited to the economic side of life. It has a tendency to favour the rich over the poor, and this can be seen in many countries today as the wealthy become relatively fewer and fewer while becoming richer and richer, while the rest of the population become relatively poorer.

In monetarist economics there is no real consideration of morals or values. It has its own ethics, but these are hidden behind a screen of arguments that 'this is what the real world is like'.

The system that has produced these results is also referred to as neoliberalism, a deliberate attempt by rich people to introduce structures that make them richer at the expense of everyone else. By clever manipulation, they have managed to deceive the majority of us that hard work and frugal living is all that is needed to give everyone the opportunity to become rich and wealthy. They have managed to convince many sensible people, especially in the USA, that cooperation is a kind of Soviet style communism that, if allowed, would turn the country into a Russian satellite state. Potent myths have been spun, and despite statistical evidence to the contrary, have become accepted as the truth. So we now believe that the Greeks and Spanish brought their economic downfall upon themselves by sitting in the shade and drinking wine. However, in *The Guardian Weekly,* February 2013, it was reported that in

2011 the average Greek worked 2,032 hours, while the average German worked only 1,413 hours, and in the Netherlands the average Dutchman worked just 1,379 hours.

Greece and Spain experienced disastrous economic collapse in 2008, and the solution that standard economics is offering them is large-scale unemployment. In the real world, when disasters happen, people clearly need to be busy clearing up the damage and rebuilding what has been destroyed. In the economic system we have today, we are told that we will become wealthy again if we make large numbers of people, especially younger people between twenty and forty, unemployed.

Insanity can be defined as having lost contact with the real world. According to this definition, monetarist and neoliberal economists are insane.

Natural economics

Economics and accounting can be used to measure performance. I often say to Permaculture students that if we can't pay the bills, it isn't Permaculture. Looking through last year's accounts should give a picture of how an enterprise fared during the year, information that can be used to plan future activities. But it will only work if the picture is a true one that takes all the factors into account. One of the problems with many corporate accounts today are that some expenses are externalised, which means they are passed on to others outside the enterprise. The best example must be nuclear energy, which is often advocated by its supporters as cheap. In fact, the costs of decommissioning power stations and dealing with the nuclear waste are not factored into the cost of the electricity, but paid for out of taxes. Factoring these costs into the price of nuclear electricity would make it ridiculously expensive. We actually have no idea how much it would cost to keep some of the nuclear wastes safe for tens of thousands of years. Had these subsidies been diverted to research and development of renewables, we would be experiencing an enormous surge of new, cheap and more efficient sources of energy.

Similarly, the cost of destroying natural resources is rarely taken into the account sheet. What is the value of a rainforest compared to the electricity generated by a large dam? It's only recently that this has started

10

to be explored. Pavan Sukhdev, former Head of Global Markets in India for Deutsche Bank, led a UN study, 'The Economics of Ecosystems and Biodiversity (TEEB)', in 2010. They concluded that the value of ecosystem services lost to the world's economy just from deforestation was somewhere between US$1.4 and US4.5 trillion every year.

A study by Trucost in 2012 concluded that the natural capital impact of the world's largest companies amounted to about US$7.3 trillion per year. If this were included in their annual accounts, most of these companies would be bankrupt.

A number of initiatives were reported in *Resurgence and Ecologist* magazine in the autumn of 2013. These approaches are now studying how to incorporate nature into the balance sheet:

- The World Bank's WAVES (Wealth Accounting and Valuation of Ecosystem Services)
- The Economics of Ecosystems
- Biodiversity for Business Coalition
- Natural Capital Declaration
- Natural or Ecological Economics are now being taught at some universities.

A cooperative keeps capital and wealth in the local economy.

Measuring economics

One measure of economic welfare used to be based on counting up the total value of goods and services produced in a country. This would then give you the Gross National Product (GNP). When the government can get the GNP to go up this is seen as a positive development.

However, it doesn't really tell you how well people are doing. If a company is busy producing something and paying lots of wages to their employees, it will register as a plus in the GNP. If one of the results of their activities is lots of pollution, that doesn't show. But as soon as health personnel get employed trying to cure those who get ill from the pollution, and government-financed clean-up operations get underway, that also counts as a rise in GNP, as everyone gets paid. Economists who use this can be accused of not discovering subtraction. Simon Kuznets, who invented this way of measuring economic activity, never intended it to be an overall indicator of human wellbeing. However, it has become the indicator that most nations now use, and is unquestioningly accepted as our way of measuring economic performance. As long as the GNP is going up, all is well.

A reaction to this arose in the 1960s and 70s with the development of a whole new range of economic indicators, such as: the 'Measure of Economic Welfare' (MEW) by James Tobin and William Nordhaus, or Japan's 'Net National Welfare' (NNW). The Society for International Development created the 'Physical Quality of Life Indicator' (PQLI), and the United Nations Environment Program developed the 'Basic Human Needs indicator' (BHN). What these had in common is a focus on wider phenomena than just money transactions. They didn't just see everything as an asset, but deducted unwanted or negative effects such as pollution, crime and war.

When we plot these over time we see that both standard GNP and these new economic indicators continued to rise in a similar way until the 1970s. Since then the alternative indicators have tended to fall, while the GNP has continued rising. The 1970s were a kind of watershed. From the austerity of the years during and following the Second World War, different forms of social democracy had led to a rising standard of living

10

for many people in the West. By the 1980s right wing governments in Britain and the United States had relaxed financial controls, freed major currencies from the Gold Standard, and the stage was set for multinational corporations to grow, and for a 'revolving door' relationship between them and governments. Monetarism or neoliberalism had set in.

Another way of measuring progress is reported from Bhutan, where the government has developed the 'Gross National Happiness' to measure prosperity. The UN has set up a panel to consider ways of replicating this idea in other countries.

Ethical banking

It has become clear to many that while mainstream banking and finance practices state that they are value-free, they actually have values such as greed built into them. As a response to this, there has been a vigorous growth of ethical and sustainable banks. In *New View* magazine, winter 2012/2013, it was reported that these banks now show higher growth rates than conventional banks. They engage with the real economy by financing social and environmental projects rather than financial speculation, and they take a prudent approach by relating lending to their capital assets. Research comparing 22 ethical banks with 28 conventional banks, conducted over a ten year period between 2002 and 2011 showed this very clearly.

In the autumn of 2008, at the height of the finance crisis, I shared an office at Cultura Bank, the ethical bank in Norway inspired by anthroposophy. Lars Hektoen, the bank's manager, told me over coffee that not only was Cultura Bank unaffected by the crisis, it was actually gaining new customers faster than ever. Because the bank only invested in sustainable projects with solid backing, and because the bank did not speculate, it was truly a sustainable bank. It had become clear to many that it was safer to put their money into Cultura rather than conventional banks.

Domestic vegetable garden ready for spring planting. Natural capital.

Biblical economics

In our western tradition we have very different ideas about economics rooted in the Bible. In Leviticus 25, God reveals a system of economics that places the individual and the family as the focus, rather than some abstract notion of 'free market forces'. This ancient Jewish tradition is concerned first and foremost with how the economy impacts on the welfare of individuals, not with how individuals impact on the welfare of the economy. Seen by God we are passers-by, visitors or strangers in the flow of time.

> The land must not be sold beyond reclaim, for the land is Mine.
> You are but strangers and residents with Me. If your kinsman is in
> straits and has to sell part of his holding, his nearest redeemer shall
> come and redeem what his kinsman has sold. (Lev. 25, 23–25)

The cycles of Shmita (every seventh year) and Joval (every fiftieth year, the origin of the word Jubilee) are calculated to both preserve the economic

10

quality of competition, which results in gaps between rich and poor, while balancing the excesses of that quality by creating new opportunities for people to break out of the cycle of poverty. Thus, loans were remitted, slaves and indentured workers redeemed and land was returned to its original owner.

> The land is a gift from God, and this gift is meant to make
> each of us a 'giver' in return. Every person must seek actively
> to cooperate with others so that the present is shared justly.
> These rules balance any tendency towards greed, and cultivate a
> sense of justice and compassion. (See article by Michael Graetz,
> *Jerusalem Report*, May 26, 2008.)

Islamic economics

In Islam, money has positive connotations; wealth is a reflection of God's favour. Money is mentioned twenty-five times in The Qur'an as God's grace and blessing, twelve times as mercy, and twelve times as a good deed. The verb 'to trade' is used positively throughout The Qur'an. The fact that the Prophet himself, the first Caliphs, great Muslim theologians and authors were well-to-do merchants conferred positive value on trade.

> Wealth and family pleasures are God's grace to mankind but
> they disguise God's test of true faith. To turn to God when one is
> needy, lonely, or sick is not a sufficient sign of true faith. Rather
> would one still remember God once one is wealthy, healthy, and
> surrounded by loved ones? To have money is a lure behind which
> lurks the ultimate test of true faith: would the joys and goods of
> the earth distract a person from remembering God? (Adapted
> from an email from Dr Ali Qleibo, lecturer on ancient classical
> civilisations at al-Quds University, Jerusalem.)

In October 2009 *The Guardian Weekly* reported that Islamic finance was one of the fastest growing sectors of the global banking industry, expanding between 15% and 20% per year. Islamic finance emerged strongly during the 1970s and is based on Sharia law, prohibiting investment involving interest, gambling and pornography. Both investor and those receiving investment must share the risks, and investments have to be backed by tangible assets. One professor of economics, Stefan Szymanski, commented that Islamic finance has much in common with ethical investment.

Different kinds of money

I have found it useful to think of money having three distinct and progressive qualities:

- Consumption money can cover daily use. Here there is a dynamic between thrift and profligacy.
- Thrift will create a surplus that will become savings. Here we think in a longer time scale than we do when we consume, when we are really concerned with gratifying needs such as hunger, thirst and shelter. Maybe we can invest in long-term projects. A loan might be necessary to develop a concrete idea. The loan must be paid back when the enterprise gets going, when it enters the consumption economy.
- If we have even more surplus money maybe we can create gift money. Banks and foundations can manage these sums. How far are we willing to give this money full freedom? Can we give it away as a grant? A new idea might need gift money to be developed. This we call risk capital. Completely free money is necessary for development, both for human beings and for enterprises.

Threefolding and peace work

The idea of 'threefolding' was presented by Rudolf Steiner as part of his lectures on anthroposophy during the last part of the First World War and the years that followed. He based his thoughts on his study of the development of European society over the preceding centuries. In England, he saw the industrial revolution as the modernisation of economic life, leading to demands for *fraternity*, the development of

10

trade unionism and labour party politics. In France under the French Revolution he saw a change in the legal life leading to demands for *equality,* and in Middle Europe (later unified to become Germany) changes in the spiritual life leading to demands for *liberty.* I like to think of this threefolding as the social application of anthroposophy.

Steiner traced how these three great ideals, of Fraternity, Equality and Liberty had been corrupted by the rise of nationalism and the development of the centralised nation state. The three-folded analysis was presented by Steiner as a way of rebuilding Europe after the disaster of the First World War, but his ideas did not gain credence, and the ideas were largely dormant until taken up by other anthroposophists.

In a lecture I attended a few years ago by the leading Israeli anthroposophist, Yishiyahu Ben Aharon, I first heard of threefolding being applied to peacework. Just as a healthy society needs a clear balance between equality in politics, fellowship in economics and freedom in culture, so does a successful peace process need to include dialogue and cooperation in all three aspects. Yishiyahu showed me how the Oslo peace process of the 1990s didn't succeed because it was only a political process, and not enough progress was made on economic and cultural co-existence.

I found an example of economics being used as a way of bringing conflicting people together in the *Jerusalem Report*, August 2011, where Shlomo Maital reports from a project at Babson College in Boston, USA. A programme called 'Bridging the Cultural Gap through Entrepreneurship' consists of bringing groups of Israeli and Palestinian students together to study business for several weeks, then returning to the Middle East to launch cooperative business ventures.

LETS and other trading systems

LETS stands for Local Exchange Trading System, or Local Energy Transfer System. Sometimes known as Green Dollar Systems, they are generally attributed to a scheme started by Michael Linton in Canada in 1983.

There are a wide variety of social purposes with such systems: from resolving local unemployment to elderly care, from mentoring kids to dealing with environmental problems. What they all have in common is to be able to operate in parallel with the conventional money system, and in being able to match otherwise unmet needs with unused resources. How can we encourage this type of economic development? Here are some questions that can serve as starting points:

- What types of products and services can be supplied?
- What is the optimum size for a pioneer enterprise?
- Do we need new business regulations?
- Do we need pioneer business zones?
- Can we organise a Pioneer Enterprise Festival to stimulate ideas and innovations?
- A Pioneer Enterprise Council might link local government, entrepreneurs, and local business people.

According to Rob Hopkins, writing in *Permaculture* magazine 77, autumn 2013, Transition Town initiatives have developed complementary currencies in several places. In Brixton they launched a 'Pay By Text' system and they allow businesses to pay their business rates in the local

10

currency and Council staff can also have their salaries paid in the currency. In Bristol, George Ferguson became the first Mayor to have his salary paid in Bristol Pounds in November 2012.

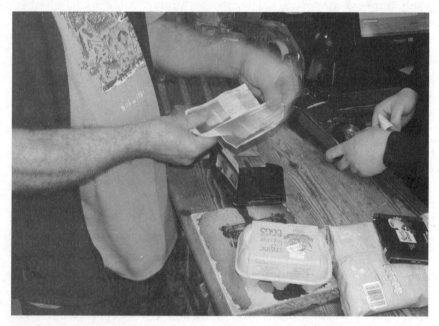

'Ecos' in use at Findhorn.

Another example of a local economy was developed in Rosenheim in Southern Germany by Christian Gelleri, a teacher at a local school. They called it the 'Chiemgauer' and after eight years it turned into one of the world's most successful alternative currencies, according to a report in *The Guardian* in October 2011. About 2,500 people use it regularly, it is accepted by over six hundred local businesses and has earned over €100,000 for local non-profit organisations.

Not surprisingly, in Spain, where the economy has more or less run out of Euros, local, alternative currencies have been booming. In September 2012 *The Guardian* reported over 325 local currencies involving tens of thousands of Spaniards. According to the article, similar projects have been developing in Greece, Portugal and other economically-troubled European countries.

Citizens' wage

In the Dauphin district in Canada there was a pilot scheme from 1974 to 1978 guaranteeing a minimum income (Mincome) for every resident. After it was over, documentation was stored in 1,800 cardboard boxes that were opened only in 2009 by a medical doctor, Evelyn Forget. Surprisingly, she found an 8.5% reduction in hospitalisation over the whole population, including those who did not receive the Mincome. Also documented were fewer work-related accidents, less violence, fewer traffic accidents and fewer psychiatric disorders.

Similar projects have been launched in Namibia, in the village of Otjivero, and in India in the Madyar Pradesh district. They show similar results, a general improvement in the quality of life, in initiatives, and a decline in social problems.

Eight forms of capital

When I was growing up the word 'capitalism' had all sorts of negative connotations amongst the radical teenage circles that I was in. I like to think we have moved on since then, and have a more nuanced view of money and capital. Ethan Roland's clear analysis, based squarely on the Permaculture principles of different kinds of assets, has been very helpful in balancing monetary and other kinds of capital and currency. I have been pleasantly surprised how well this idea has been received in the Permaculture Design Courses that I teach. The following is based on an article in 'Eight Forms of Capital' by Ethan C. Roland of Appleseed Permaculture, USA (*Permaculture* magazine 68, summer 2011.)

Ethan's starting point is a definition of capital as 'wealth in the form of money or other assets', which opens up a whole new spectrum of ideas under 'other assets'. He sketches out the following:

10

- **Financial capital** has as its currency *money,* leading to financial assets and securities.

- **Social capital** has as its currency *connections,* leading to relationships and influence.
- **Material capital** has as its currency *materials and natural resources,* leading to tools, buildings and infrastructure.
- **Living capital** has as its currency *carbon, nitrogen and water,* leading to soil, living organisms and ecosystems.
- **Intellectual capital** has as its currency *ideas and knowledge,* leading to words, images and 'intellectual property'.
- **Experiential capital** has as its currency *action,* leading to embodied experience and wisdom.
- **Cultural capital** has as its currency *song, story and ritual,* leading to community.
- **Spiritual capital** has as its currency *prayer, intention and faith,* leading to spiritual attainment.

Using these ideas as starting points, we can explore the relationships between paid work, volunteer work and activities that are done for pleasure, combining productive end results with positive social experiences. At the end of the article Ethan warns us that there might be a danger that we commodify ecosystem services, spirituality and culture, and that 'Time and Labour' are not clearly factored in. As with so many things in Permaculture, the analysis is a starting point for future exploration rather than a given set of 'truths'. In an Ecovillage setting these eight forms of capital could be the basis of a study-group working to further develop indicators to show how the Ecovillage is developing.

The amphitheatre at Damanhur, filled with the International Communal Studies Association 2007 conference.

Community profile

Damanhur

Damanhur is a Federation of Spiritual Communities located in northern Italy between Turin and Aosta, an area named Valchiusella. Damanhur was founded in 1975 under the inspiration of Oberto Airaudi (1950–2013), known as Falco (Hawk), and now comprises about six hundred residents.

Every year thousands of people visit Damanhur to try out the social model, study the philosophy and to meditate in the Temples of Humankind, that great underground construction excavated by hand by the citizens of Damanhur and which many have called the 'Eighth Wonder of the World'. The Temple Halls are an underground work of art, a subterranean cathedral created entirely by hand and dedicated to the divine nature of humanity. It is a great three-dimensional book that recounts the history of humankind through all the art forms, a path of re-awakening to the Divine inside and outside of ourselves.

The Damanhurian complementary monetary system has been created

Supermarket at Damanhur selling vitalised 'selfic' food.

to give back to money its original meaning: to be a means to facilitate change based upon an agreement between the parties. For this reason it is called Credito, to remind us that money is only a tool through which one gives trust. Thanks to this monetary system, the Damanhurians want to raise the quality of money, by not considering it an end in itself but as a functional tool for exchange between people who share values.

The use of the Credito allows people to see themselves as part of the cultural, social, economic and ethical values linked to the sustainability of the planet. It encourages respect for human beings and for all living creatures.

A wide network of local producers and consumers, including a hundred businesses and over two thousand people, have chosen to use the Credito through a system of agreements. In this way, the Credito encourages an economic and social revitalisation of Valchiusella because it facilitates keeping capital inside the area so that it can be re-invested for the benefit of the local economy.

Linked to the project of the Credito and as a complement to it, there is DES, a system of social loans that accepts deposits and issues financing. DES arose from the idea of a finance service to start up development projects with ethical and social aims. It is possible to participate in these projects by opening a savings account, available to members, with advantageous rates of interest.

In purely technical terms, the Credito is a unit of a working account, used within a predetermined and predefined circulation. Today the Credito has the same value as the Euro.

For more information, see: http://www.damanhur.org

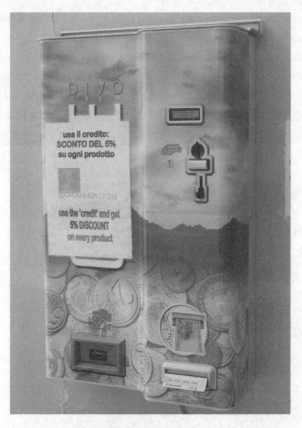

Buying things at Damanhur with the local currency 'Creditos' gives a discount.

10

Tips for facilitators and groups

Create an alternative currency

One of my great tools is the 'Post It Note'. Just coloured pieces of paper with a strip of glue on one side. You can paste them up on walls or windows or any surface. They come in varying sizes and colours.

This exercise entails asking each participant to think of two to four things that they can offer others, and two to four things they need. These can be products or services. A gardener can offer boxes of vegetables, a bike mechanic can offer to fix bikes, a masseur can offer a massage, a cook can offer a meal. Things that they need can be written on red notes, and things they offer on green ones. Ask them to keep it simple and realistic. Remember to ask them to put their names, and if it's a new group, their telephone numbers on each note.

Creating an alternative currency. From a Permaculture Design Course in Norway, 2013.

It's good to give them ten to fifteen minutes, but no longer; keep it short and simple. Once they have written, ask them to paste their notes up on the wall, keeping red and green separate.

Once everyone has written, it's time for a break, and you can ask two volunteers to start matching up the notes, red and green. Anyone can watch and help. Once everyone has a cup of whatever, tea or coffee, get the group together and explain that you have now created the basis for a trading system. All they have to do is agree on a common currency, it could be time (ecohours) or an agreed payment of something symbolic (ecobeads). This is the beginning of real economics, and a break to give them time to trade would be appropriate at this point. If the seminar stretches over several days, it would be good to give the group additional time to develop this trading system.

10

Ritual and celebration helps to weld a group together.

Chapter 11. How Groups Develop

Any group of people getting together to form community or even an association that is to last over time, creates an organic social form. These forms have a life of their own, and after a while we can look back and trace the biography of the group, and find that each one has a distinct unique history. Even though individuals may come and go, the group has its own identity. Just like individual people, groups go through developmental stages, and we can trace patterns, such as youth, maturity and old age. Transitioning from stage to stage is often dramatic, not unlike the changes we go through when we experience puberty. Understanding these patterns can help us through those hard times.

The Permaculture, Transition Town and Ecovillage movements have developed a number of useful tools to help groups through the hard times, and we will look at some of these tools in this chapter.

Course lifecycles

A design course lasting a week or more also has a lifecycle. The group behaves and feels very differently at the end of the course than it did at the beginning. In order to get off to a good start, there are many different tools we might use. Some of these have been gathered at the end of preceding chapters of the book.

We might want to begin with a series of statements and questions, to be answered as a go around (see below).

- Tell the group about a meaningful experience that you have had in connection with community.
- What aspect of community brought you to this course now?

- Are you in any kind of community or group that is meaningful to you now? Please describe it.
- Are there any negative aspects of community you would like to talk about with the group?
- What particular aspects of community would you like to explore during this course?

Every group develops its own culture, something that contributes to the group identity. Being aware of this, and consciously developing it, helps the group to establish a safe and comfortable place for each participant. We can brainstorm attitudes, behaviour and guidelines to secure respect, positive attitudes and mutual support within the group. These can be written on a mind map on a large piece of paper which can be fixed to the wall, allowing us to refer to them should any kind of conflict arise.

Many people are apprehensive at the beginning of a course. New people, unknown challenges ahead, and negative experiences in the past may create insecurity. One way of overcoming this is to ask everyone to write down two or three fears they have on a piece of paper, without

Creating trust in a group is the first priority, whether it's a learning group or a decision-making group. Here from a design course in Latvia in 2013.

putting their name on it. These can be collected in a box, shaken well, and then the box is passed around the group, each person taking out one slip of paper and reading what is written aloud. This is very helpful for the facilitator who can give reassurance in sensitive areas. At a recent design course I taught, several participants voiced a concern about the individual design projects, fearing that they may not be able to achieve a high enough standard. I made extra sure to explain the design criteria carefully, give sufficient time for them to work on their projects, and positive encouragement and advice along the way. Everyone presented a satisfactory design at the end of the course.

Making decisions

A group making decisions has to operate in a number of different areas. Making these explicit may help the group to move forward when it gets stuck. This is not a linear progression, and a mind map with the aim in the middle might help the group to define where they are at any given moment. You might like to focus on one or more of the following:

- *The aim.* What are we striving for? Where are we going? When everyone shares the same aim, decisions are much easier to arrive at, but if some members are there for completely different reasons, disagreements may arise.
- *The action-plan.* What needs and resources do we have at our disposal? What are we actually going to do? Do we need to divide into workgroups or delegate some things to outside people or groups?
- *The facts.* What information do we need? Do we have the information we need?
- *The individuals.* Is everyone able to contribute? How can each person's skills and interests be incorporated into the plan?
- *The development.* How far have we come in our journey to fulfil the aims? It's often very helpful to stop the flow of discussion occasionally to see where we are.

11

- **The achievement.** When we finally come to a decision, we may ask ourselves if we really have achieved our aim. This might be a kind of evaluation.

Tools for better meetings

There are several tools that we can bring with us to meetings, in order to make them better and more enjoyable.

Listening exercise

Divide into pairs, each one takes a few minutes to talk about what is going on, how they are feeling. The partner just listens, does not ask questions or make comments. Then they switch roles.

Go arounds

This is a standard meeting and seminar tool, used on many occasions. It gives each person a chance to talk and be heard, and breaks up otherwise long and sometimes confused or rambling sessions. A go around can be used for pretty much anything: how people feel, their reaction to a topic, a check-in first thing in the morning, and a quick evaluation at the end of a session or a day. Sometimes go arounds are called 'creeping death', as the last few participants see their turn come slowly creeping up on them, and they don't know what to say.

Popcorn

Like a go around, but instead of the structured round, people elect to speak when they are ready. Again, it can be used for anything to get the energy in the group to change. In my experience as the group becomes more comfortable with itself, there are more popcorns than go arounds, but sometimes popcorns spontaneously become go arounds.

Both popcorns and go arounds are useful ways to begin and end meetings. At the beginning, a quick 'weather check' of how people are feeling may reveal sensitivities in the group that the facilitator should be aware of, and to end a session or a day with a quick evaluation is a useful way for the facilitator to assess how things went and adjust accordingly for the next day or future meetings.

Everyone speaks first

Occasionally go arounds or popcorns are interrupted by others making comments and asking questions. It's quite useful for the facilitator to check these, pointing out that the group is going to let each person speak before opening up to a discussion, if that is what is needed for the next step. This is important in order to let the shyer participants have a chance to talk, and not to let the more vociferous members of the group dominate the meeting.

Speaking from yourself

Often people hide their own feelings by trying to make them universal, saying things like: 'It's not such a good idea to …' when they really mean, 'I don't think that's a very good idea.' A facilitator might ask everyone to use the 'I' form when speaking, instead of 'someone', 'you' or 'one'.

Consensus

I worked for a while as a Trade Union branch secretary in England, leading meetings every month with a rule-book. Everything was decided democratically, with motions forwarded, seconded, discussed and then put to a vote. We had a similar, if less formal, decision-making system in the kibbutz I lived in. Democracy is a good system, and obviously much better than a dictatorship of the strong, but it does have the built-in problem of creating winners and losers.

11

It's really important to have a wide range of emotions, and give them free rein.

Consensus was developed to overcome this flaw, and has been used for many decades and now has a good tradition and its own training systems. One important point is that it's not just talking and talking until everyone agrees (or not!) but is based on rules and training, which are really important to make it work well. Here are a number of points that should be borne in mind about consensus decision-making, but they are not a substitute for real training. Any group wanting to use consensus should study the system and are advised to begin with a trained facilitator.

Why consensus

There are many reasons for moving "beyond democracy" and you will have to assess what is right for the group that you are participating in. Here are some of the reasons why it might be appropriate to make such a move:

- Majority voting creates losers and encourages conflict.
- Consensus gives each person's voice a chance to be heard and requires agreement from everyone involved.
- Consensus includes everyone, helps create a common vision and encourages strength in a group.

Consensus does have its weak points. It is a slower process and more complicated, and any one person can block a decision, which may lead to resentment and frustration in others. It can also lead to no decision, which may undermine the group work. Consensus also demands a willingness from everyone in the group to work with it.

In some instances I have come across what I might call 'consensus lite', a form of consensus where a vote might be taken if things take too long and are too complicated. This is not always successful, but works well if the group is friendly and trusting and don't have any deep-seated conflicts or controversial decisions to be made.

The process of consensus

Experience has shown that successful consensus requires training, and the inclusion of a facilitator at the beginning. The following is a list of what could be the typical sequence in a consensus decision-making process:

- Gather questions and make an agenda, with priorities and timing.
- Talk the issue through as a group. Try and see it from various angles.
- Make proposals.
- Take the proposals and work on them, changing the wording so that everyone gets included.
- Can we all agree? If not, are those who don't agree willing to support? Different ways of expressing this can be used, hand signals, cards, or even moving around the room.
- Individuals can 'agree to disagree' and support the proposal in order to show solidarity.
- Anyone seriously disagreeing with the proposal can choose to 'block' in which case the proposal falls. Amendments to the proposal can be suggested to overcome the blocking, but if this fails, the proposal remains blocked.
- In practice, it has been found that very few proposals stay blocked, and that the blocking option is pretty serious, and is only seldom

11

used. If anyone is in a blocking position more than once or twice, it opens a question how well they fit in with the group.

Consensus tasks

There are a number of defined tasks that have been developed to make consensus flow easier, and any group wanting to work with consensus should train in these tasks so they can go in rotation.

- **Facilitator.** This is the discussion leader, directing the conversation. This person frames the proposals, and tries to keep an eye on where the group finds itself at any given time. You might like to refer back to 'Making decisions' and 'Tools for better meetings' (see above, p. 221 and p.222).
- **Minute taker.** Writes down the proposals and the changes in the proposals that the group decides upon.
- **Timekeeper.** This person keeps an eye on the time, keeping track of time agreements. This is especially important so that meetings don't drag on forever. This person can interrupt at any time with such information as: 'We agreed that we would spend forty minutes on this point. There are now five minutes left and we should have a summing up before we go on to the next point.' Should the group decide to spend extra time on a point, the Timekeeper may ask which of the other remaining points should be held over to a subsequent meeting.
- **Heart watcher.** This person keeps an eye, or heart, on the feelings that emerge in the group as the discussion continues. It's quite usual for this person to report back at the end of the meeting, giving a little evaluation. The heart watcher can also interrupt at any point, asking for a short break, or a clarification, or that the group reappraises its language and/or behaviour.

Time for personal talk is really important for a group to develop.

Open Space Technology

This was started in the early 1980s by Harrison Owen and David Belisle, and developed over a few years at annual Organisational Transformation (OT) conferences held in the USA. At first they used the traditional conference format but after three years Owen did not relish another year of work to manage all the details, and claims to have invented the Open Space Technology over a few martinis in the bar. He never trademarked or patented 'Open Space' in any way. He said it could be practised freely by anyone with a good head and good heart.

I have come across more and more groups using this meeting form in various ways over the last few years. Maybe its main value lies in opening up the ways we manage meetings, giving free rein to creativity and experimentation.

11

Typical meeting process

At the beginning the participants sit in a circle, or in concentric circles for large groups. Open Space has used variations on this in small groups of ten to fifteen, but it was first developed for much larger groups of 300 to 2,000 people or more.

The facilitator will greet those present and briefly state the theme of their gathering. Then the facilitator will invite all participants to spend the next ten minutes in thinking through and identifying any issue or opportunity related to the theme. When the facilitator announces time is up, any participants willing to raise a topic will come to the centre of the circle, write a short description, just a few words, on a sheet of paper and announce it to everyone. The person who has called out this issue or opportunity then posts the paper in an area of wall designated for the agenda. That wall becomes the agenda for the meeting. When all those who have topics they want to share have announced and posted their topic, everyone is invited to gather round the topic that most interests them. There is no commitment, each person can stay with a group or move on, according to their interest, and how they feel they are contributing or not.

Sessions typically last for an hour and a half; the whole gathering usually lasts from half a day up to two days. The opening and agenda creation session lasts about an hour.

After the opening and agenda creation, the individual groups go to work. The participants organise each session as they go, decide which session they want to attend, and may switch to another one at any time. This supports different styles of participation as many people like to sample before committing to a group, others may be looking for the most productive sessions, while yet others are hoping to pinpoint discussion on an issue. All discussion reports are compiled in a document on site and sent to participants shortly after.

In this way, Open Space Technology begins without any pre-determined agenda, but work is directed by a theme or purpose that is carefully articulated by leaders, in advance of the meeting. The organisers outline in advance a schedule of break times and spaces. The combination of clear purpose and ample break facilities directly supports the process of self-organisation by the meeting participants themselves.

Small groups might create agendas of only a few issues. Very large groups can generate many sessions running concurrently over the course of a day, and longer meetings may establish priorities and set up working-groups for follow-up.

Conflict resolution

Conflict can be seen as a disease, with a diagnosis and a prognosis. If we think of it as a disease, we might look for some social bacteria, the Rumour bacteria, the Fixation bacteria and the Persuasion bacteria. Our first approach, following this analogy, might be a conflict diagnosis. In order to make this easier we might ask a number of questions:

- What is the basic issue?
- What has happened up to now?
- Who is involved in this conflict?
- What are the personal and formal relationships between the partners?
- Can these partners solve the conflict with their present attitudes?
- Is there something underlying this conflict?

It is often easier to have a third party come in and help with such a process. This facilitator or mediator has several roles and may do one or more of the following:

- Make observations as objectively as possible.
- Be neutral and not moralise.
- Try to see the situation from a distance.
- Be accepted by all parties.
- Develop trust.
- Awaken the desire to take responsibility.
- Raise awareness.

11

If conflict is a social disease, or a form of pollution, we might consider it in an entirely different way. First of all, it might not be a wholly negative thing. Disease and pollution are, like pain, signals we need to respond to. If we regard them as symptoms, we might see them as useful signals telling us that something is wrong and that we need to find out what is wrong and address that issue.

Just as disease can be seen as a signal for improving health, and pollution as resources looking for a use, social conflict might be regarded as an opportunity for self-growth. It might be helpful to be aware of the following:

- My ideas, where do they come from, are they my own?
- My decisions, how do I make them, if at all?
- My way of talking, how do I say what I say in any given situation?
- My actions, how do they affect my surroundings?
- My strivings, can I lift myself up over the everyday, where am I going?
- My capacity to learn from life, from my own and others' experience?
- Being aware of the part of my individuality that can pose these questions.

Qualities that we could develop from this kind of training would include listening with loving interest, creating a common picture or a common point of departure, and the ability to act in freedom. We need to develop the capacity to observe, to evaluate, and be able to take free actions. All this demands work upon the self, not in isolation, but in the context of social interaction. These are not lists to be followed slavishly, or memorised by heart, but rather ideas and processes that can be observed, understood and put into practice.

Active hope

This exercise can be done either individually in a short time, or in groups, spending as much time on it as they want. The exercise is an ascending spiral, itself a powerful Permaculture pattern, and is ideal for an Ecovillage to come closer together and understand the connections between individuals and the group.

- *Gratitude.* Start a sentence with 'For supporting me to live, I give thanks to…' Complete the sentence to give a starting point for giving something in return, generating motivation for positive actions.
- *Honour our Pain for the World.* Start a sentence with 'Looking into the future we're heading into, concerns I have include …' Again, complete the sentence to create motivation and a sense of urgency.
- *Seeing with New Eyes.* Start a sentence with 'Something that inspires me is …' Hopefully, once this sentence is completed we should find ourselves locating a response to the first two steps.
- *Going Forth.* Start a sentence with 'Something I'd love to do to make a difference is …' Once this sentence is completed, try another one: 'A step towards this I will make in the next seven days is …' By now you should have a plan of action. If doing this in a group, a whole list of things you might want to do.

The spiral nature of this exercise comes when it is repeated at regular intervals, each time building upon the work done before, inspired by the actions that have been done since. In an Ecovillage group, this would be a powerful tool. Use it regularly, once a week or once a month. (Based on an article by Chris Johnstone *in Resurgence and Ecologist* magazine, 283, March/April 2014. See also Chris Johnstone, together with Joanna Macy, *Active Hope: How to Face the Mess We're in Without Going Crazy.*)

11

Rudolf Steiner's words

Having lived for several years in a Camphill village, I would like to complete this chapter on group processes by sharing some thoughts from Rudolf Steiner that have inspired Camphill. The following words were often repeated: at the beginning of meetings, at the beginning of the day, and sometimes even spontaneously in a meeting or a lecture.

> Health and wholesomeness only come
> when in the mirror of the soul of man
> the whole community takes shape;
> And in the community lives
> the strength of every single soul.

These words are useful in reminding us that there is a non-physical quality about a group of people when they are working together in community. All the different group processes described through this chapter are merely tools to enhance that quality.

The lecture, given by one person to a large group, is a classic educational tool.

The shop at Svanholm.

Community profile

Svanholm

Founded in 1978, Svanholm is a secular, rural community in Denmark, with a population in 2012 of about 140 people, of which there were about eighty adults and forty-five children.

From 1978 until 2003 they had complete economic fellowship, in which all incomes went to the collective and the individual members received pocket money. Since 2003 they have experimented with members receiving pocket money related to income, set at 25%. The rest goes into the collective, but members need to pay for slightly more of their private consumption than previously.

The Svanholm Estate comprises 400 hectares of land where they practise organic farming with a strong and well-established commitment to the environment. They produce about 300 tons of grain and 200 tons of root vegetables annually. They have about 100 cows for milk and meat, and 200 lambs, plus extensive vegetable production and fruit trees. Committees or work groups deal with economics, work coordination, investments, new members, guests and allocation of housing.

11

A lengthy membership process establishes trust between the members, and they have a strongly child-centred culture. Svanholm is very established and very stable.

For more information: http://www.svanholm.dk/index.php?id=73

Large community gardens help feed the residents at Svanholm.

Tips for facilitators and groups

Evaluations

At the end of any meeting session or course, it is extremely helpful to make an evaluation. This helps the group look at itself as a group, and is also one of my tools for improving teaching skills and leadership meeting techniques.

A short version is just to go round the group and ask:

- What do you think the group is doing well?
- What do you think could be improved?

This is not an opening for comments and discussions, but a chance for each person to voice their own opinions. I find it useful not to comment on any person's evaluation, but to simply thank them for what they said and note it for future reference. Even when I receive direct criticism I refrain from answering defensively, again just thanking them for their comments and noting down what was said.

A longer evaluation might entail a series of questions, and then it might be better to have a written form, otherwise it takes a long time to get round the whole group. Personally I have never found written evaluations very much fun to fill out, and hesitate to inflict this on others. However, in case you find it important to have a written version the questions might look somewhat like this:

- Did you hear something today that stood out for you?
- What new ideas or perspectives did you hear?
- What resonated with the whole group; what themes came up in the conversations?
- Do you want to continue this exploration of community?
- Are there any specific actions you would want to take on after what you heard today?

In addition you might want to list some logistical aspects for future organisation:

- How was the food?
- How was the accommodation?
- How was the teaching space?

11

Chapter 12. Crises and Solutions

Today we face a number of serious, existential challenges. This closing chapter looks briefly at some of them, and how we can turn them to our advantage. This is one of the classic Permaculture principles, turning problems into solutions, and as Ecovillage designers we would do well to use that principle.

Mass extinctions in the past

At school I was taught that the age of the dinosaurs came to a dramatic end, ushering in the age of mammals. Research since then has revealed that this was not the only extinction.

- The Ordovician-Silurian era, when most life was in the seas, experienced an extinction about 443 million years ago when 85% of marine species died out.
- During the Late Devonian period, about 359 million years ago, roughly 75% of species went extinct over a period of several million years. Geologists think that this may have been due to asteroid impacts.
- The biggest mass extinction the geologists have found occurred during the Permian period, about 252 million years ago. Possibly due to volcanic activity, they think that about 96% of marine species died out.
- During the Triassic-Jurassic, 200 million years ago, about half of all species died out over a period of several million years, though most plants managed to survive.
- The last extinction occurred 65 million years ago during the

Cretaceous-Tertiary period when the dinosaurs became extinct. Some geologists believe it was due to a massive asteroid impact.

It seems that there has been a slow general extinction rate, fairly well balanced by the emergence of new species due to natural causes during the time between these major extinctions. Today, however, the rate of extinctions is several times faster than that background exchange. According to some biologists' predictions, half the known species we share the planet with may be extinct by 2050. Whether this is directly a result of human activity is still hard to prove, but it does give us some food for thought. We might ask ourselves some questions.

- How do we prepare for a world that has significantly fewer species?
- Is there something we can do to preserve as many as possible?
- What kind of opportunities may arise for new species to emerge?

Climate change

In May 2013 *The Guardian Weekly* reported that: 'For the first time in human history, the concentration of climate-warming carbon dioxide in the atmosphere … passed the milestone level of 400 parts per million.'

There is no doubt that our human activities, especially the burning of fossil fuels, have contributed to this situation. Many people with strong vested interests in maintaining the status quo, have denied our involvement, but over the past year, scientific evidence has proved them wrong. To maintain climate change denial today is really irresponsible and can only be regarded as burying one's head in the sand.

The International Panel on Climate Change (IPCC) has issued a number of reports, the latest one as I write, in April 2014. Produced by 1,250 international experts and endorsed by 194 governments, this latest document was summed up by Leo Hickman writing in *The Guardian Weekly*: 'Climate change is real. We are to blame. It will get worse if we fail to act. The solutions are available and affordable.'

Climate change is already upon us, and the main points of the latest report include the possibility of threats to our food, to our security and the very real problem of inequality. No one is safe from this, and the future looks hard, but not hopeless. This latest statement is of real importance, because hopelessness is as damaging as denial. The whole thrust of Permaculture, and its manifestation in Ecovillages, is about doing what we can do, with what we have.

Peak Oil

This term 'Peak Oil' comes from the American geologist M. King Hubbert who defined *Hubbert's Peak of Oil Production* as the top point in oil extraction from one or more wells or oilfields. It refers to the top of the bell curve of oil extraction. Researching this in the 1950s, Hubbert found that this pattern holds true for individual oil wells, for oil fields and for global oil extraction. His predictions were ridiculed at the time, but fifty years later, they have been totally verified by oil statistics.

It's important here to distinguish between oil reserves and the rate of extraction. There may be enormous reserves, but only a limited rate of extraction. What makes 'Peak Oil' so important is that we seem to be reaching the top of the global bell curve right now, either during the last ten years, or during the next ten years. According to Hubbert's analysis, the bell curve is always roughly symmetrical, so if oil began to be pumped out of the ground at the end of the nineteenth century, and the peak is about now, we can expect the curve to near its zero about a hundred years from now. In 2007 annual production was 85 million barrels per day. When annual consumption becomes more than annual production, prices will go up, something which will have wide ramifications upon an economy in which the technology is so totally dependent upon oil.

Wood will become increasingly important as a renewable fuel. This from Twin Oaks community in the USA.

Our economic system is based upon the idea of exponential, eternal growth, and the idea of 'Peak Oil' represents an enormous existential threat to this system, which is one of the reasons that most mainstream economists and politicians can't deal with it. As long ago as 2005, the Worldwatch Institute recorded a reduction of oil production in 33 of the 48 world's largest oil-producing nations. Here in Norway we passed our peak of production in the early 2000s, and ten years later production is down 45% already. We started pumping oil out of the North Sea around 1970, so we can expect that our oil extraction and revenues will be insignificant around 2030, only fifteen years from now. There is a deafening silence about this in our media, despite solid research having been done on this by many respectable institutes.

One of the ways that economists and politicians try to pull the wool over our eyes is to invoke new technology, such as tar sands and fracking. Here it's really important to bear in mind that Hubbert was not writing about how much oil there is or is not on the planet, he was researching the rate of oil extraction. Obviously one extracts the easiest and cheapest oil first, and this will leave the more difficult and dearer oil to be extracted later. Oil is a form of energy (that's why we extract it!), and we use energy to extract it. When we get to a situation where we use more than one barrel of oil energy to extract one barrel of oil, we are clearly entering 'The Age of Stupid'.

Similarly, certain kinds of extraction entail enormous environmental damage. Fracking, known as hydraulic fracturing, uses explosives to open cracks in oil- and gas-bearing shale rocks and forcing fluid through these cracks, often several kilometres below the surface. This fluid is made up of water and sand, with about a 2% mix of several hundred highly toxic chemicals, including mercury, lead, uranium, benzene, toluene, ethylbenzene, xylene and methanol. These chemicals inevitably seep into ground water and contaminate aquifers, and many of them are known to be carcinogenic.

Monitoring the environmental impact of extraction techniques should give us some measure of the environmental cost. When that cost becomes too high, sensible and responsible decisions will involve deciding not to extract, even if the oil reserves are there.

In clear statistical terms we need to address the issue of 'Energy Returned On Energy Invested' known by its great acronym EROEI. When the first oil was being pumped a hundred or more years ago, EROEI was about 1,200, really high, making oil extraordinarily cheap. Today, conventional oil extraction lies between 10 and 20, while tar sands and shale oil are only 2 to 5. Interestingly, ethanol and biodiesel rarely break even, so are hardly worth chasing unless some radically new technology can be found.

There is no need for immediate panic; we have many years of oil still left on our planet, but we need to use these resources over the coming years to design a future independent of oil. Not to do so is clearly ensuring

12

Restoring terraces at Torre Superiore Ecovillage in Italy. A long-term design solution.

catastrophe upon our grandchildren. Buckminster Fuller saw this forty years ago and wrote in his book *Operating Manual for Spaceship Earth*:

> We cannot afford to expend our fossil fuels faster than we are 'recharging our battery', which means precisely the rate at which the fossil fuels are being continuously deposited within Earth's spherical crust.

> The fossil fuel deposits of our Spaceship Earth correspond to our automobile's storage battery which must be conserved to turn over our main engine's self-starter. Thereafter, our 'main engine,' the life regenerating processes, must operate exclusively on our vast daily energy income from the powers of wind, tide, water, and the direct Sun radiation energy.

Design solutions

Extinction, Climate Change, Peak Oil and depletion of resources all sound like bad news, and there are plenty of people banging the doomsday drum. However, by using the Permaculture principle of turning a problem into a solution, we have good reason to be optimistic about the future.

Permaculture design gives us tools that we can use for personal, social and technological change, transforming existing settlements and designing Ecovillages into sustainable communities where we can live high quality lives with a rich and rewarding culture. According to another Permaculture principle, we can only do what we can do, and one thing we can do is to work within the community where we find ourselves. Transforming that community into a sustainable and resilient community of the future will be a thrilling and gratifying social experience.

Holistic design implies bringing in all values, and the ability to listen. Because we are social animals, we do most things better when we work together with others. Our design process will always be enhanced by having several people working together. An Ecovillage or Transition Town group is a great starting point for exploring how we want to live together, and using the design tools suggested in this book, transforming the group ideas into a reality.

Permaculture is planning. We follow the path of creation, from an idea to action. This process is what is important. This is what we Permaculturists are concerned with. If we regard it as a linear process, we could set it out as follows:

- *Idea, observation and thinking:* This is mostly carried out in our heads, or in discussion.
- *Planning:* This is mostly done on paper.
- *Action:* This is largely carried out with tools in the physical world.
- *Result:* Things created in the world.

Our scientific methodology today is still largely based on analytical and reductionist thinking that arose around three hundred years ago. Descartes

12

Presenting a design at a Permaculture Design Course in Norway in 2013.

asserted that only those things that we can measure are important for science, and since his day we have reduced our world to increasingly smaller bits in order to look at each bit more closely. This has been fine for inventing steam engines, mass production and atomic weapons, but clearly has not resulted in better quality societies, contented people or cooperation with the natural world. Ecology as an accepted science arose in the 1960s, and to a certain extent broke out of this narrow-minded reductionism in order to look at things in their context. But even ecology is largely based on biology, geography and other natural sciences without really questioning their philosophical foundations.

As long ago as the 1940s, Arthur Koestler wrote:

After all it is only three centuries since God became a mathematician and we have plenty of time before us for other transformations. The monopoly of quantitative measurements is drawing to its close, but already new principles of explanation begin to emerge. (*The Yogi and the Commissar*, page 206)

We need a new conscious holistic methodology; a methodology that can combine physical processes with the fundamental patterns that we see in time and space. We need to bring in our own human personalities as observers of these processes. We need a methodology that recognises that there is more to our world than just its physical components; a methodology that is not only quantitative, but also qualitative.

In addition to analysing and measuring, we need to look at the world in other ways, integrating context and metamorphoses, over time and space. We need to look at processes and relationships, at change throughout life cycles.

In this design process we train ourselves. The observer becomes part of the work, part of the process. We use not only the microscope and the measuring tape, but also our human faculties and skills. We use intuition, imagination, creativity, fantasy, artistic skills and spirituality. We grow as human beings. We understand not only more about the world; we understand more about ourselves.

It is important that Permaculture is based upon strong scientific principles, and that it does not accept reductionism unquestioningly. It has become clear to me that for the last few centuries we have been working with an incomplete picture of the world. A better understanding of the world could help us to create a better world.

We can design ourselves out of the mess we have made, but only with the right tools. Some good tools that we might use are Ecovillage Auditing, Environmental Footprint Analysis and the design system developed by The Pachamama Alliance Symposium.

Ecovillage profile auditing

This was developed during the first years that the Global Ecovillage Network was establishing itself, and we needed an indicator to help Ecovillages check their rate of progress. Activity was divided into four areas, and these areas are still the basic frameworks for the Ecovillage Design Education course (EDE).

- *Air: Culture.* This includes cultural events, rituals, customs, communal space, openness and service to others.
- *Earth: Physical environment.* This covers food, buildings, restoration of nature, solid waste, place and size.
- *Water: Physical infrastructure.* Here we find energy, water sources, transport, information and wastewater.
- *Fire: Social and Economic environment.* These are decision-making processes, outreach, health, economics, services, mix of dwellings and work places.

A summary of these four areas leads us to the core of the exercise: Quintessentia: Community. This is a product of balance of the previous four, with two additions: vision and commitment.

For a quick check, points could be assessed for each of the four themes in each of the four elements, between 0 and 4, 0 meaning that there was no interest in this theme, and 4 meaning that it was well developed and functioning. Ecovillages scoring 64 were obviously all the way there, a shining example of Ecovillage living. If you score 0, it would seem you still have a long way to go. However, it wasn't meant to be used to score Ecovillages against each other, but to assess progress over time. The value of this indicator is highest if the audit is carried out every year and the scores compared to the year before, giving an indication of how well the Ecovillage is doing as it develops and grows.

Culture/Spirituality – Air

- Creativity, arts, personal development.
- Rituals, celebrations and cultural diversity.
- A new holistic, circulatory worldview.
- A process towards peace, love and global consciousness.

Infrastructure – Water

- Water care in village and bio-region.
- Integrated renewable energy systems.

- Reduction in transport.
- Access to phone, fax and e-mail.

Ecology – Earth

- Bioregional organic food supply.
- Ecological building.
- Life cycle analysis of products.
- Renovation of nature.

Social Structure – Fire

- Decision making between participants.
- Sustainable economics.
- Preventive and general health care available to all.
- Teaching and outreach.

Sharing abundance. From an Ecovillage seminar held at Hurdal Ecovillage in Norway in 2005.

12

Environmental footprint analysis

In 1996 Mathis Wackernagel and William Rees published a book called *Our Ecological Footprint.* Working at the University of British Columbia, they had developed a method of analysing our ecological impact by translating every resource use into how much land it required. In this way they had developed a new, understandable and useful indicator.

Ecological footprint analysis compares human demand on nature with the biosphere's ability to regenerate resources and provide services. It does this by assessing the biologically productive land and marine area required to produce the resources a population consumes and to absorb the corresponding waste, using prevailing technology. This approach can also be applied to an activity such as the manufacturing of a product or driving of a car. The consumption of energy, biomass, building materials, water and other resources are converted into a normalised measure of land area called 'global hectares' (gha).

Ecological footprints may be used to argue that many current lifestyles are not sustainable. Such a global comparison also clearly shows the inequalities of resource use on this planet at the beginning of the twenty-first century.

The UK's average ecological footprint is 5.45 global hectares per capita (gha) with variations between regions ranging from 4.80 gha (Wales) to 5.56 gha (East England). Two recent studies have examined relatively low-impact small communities:

- BedZED, a 96-home mixed-income housing development in South London, was designed by Bill Dunster Architects and others. Despite being populated by relatively 'mainstream' home-buyers, BedZED was found to have a footprint of 3.20 gha due to on-site renewable energy production, energy-efficient architecture, and an extensive green lifestyles programme that included on-site London's first carsharing club. The report did not measure the added footprint of the 15,000 visitors who have toured BedZED since its completion in 2002.
- Findhorn Ecovillage had a total footprint of 2.56 gha, including

both the many guests and visitors who travel to the community to undertake residential courses there and the nearby campus of Cluny Hill College. However, the residents alone have a footprint of 2.71 gha, a little over half the UK national average and one of the lowest ecological footprints of any community measured so far in the industrialised world.

Ecological footprinting is now widely used around the globe as an indicator of environmental sustainability. It can be used to measure and manage the use of resources throughout the economy. It can be used to explore the sustainability of individual lifestyles, goods and services, organisations, industry sectors, neighbourhoods, cities, regions and nations.

The Pachamama Alliance Symposium

This was developed in South America as a means of shifting our culture from one based on consumption and destruction to one based on justice, sustainability and fulfilment. The aim is to bring forth an environmentally sustainable, spiritually fulfilling and socially just human presence on the planet. Again, in true Permaculture style, this begins by asking a series of questions.

- *Where are we?* We start by simply looking at how we're doing, what is our current 'dream' creating?
- *How did we get here?* We offer a fresh perspective on how we got into this situation – the root causes.
- *What's possible for the future?* What is possible now, and what is actually already emerging in the world?
- *Where do we go from here?* Where do you fit in? How does this affect you, your life, your community and the human community?

12

From Zone 00 to Zone 5

I would like to give the last word to Eileen Caddy, founder of the Findhorn community and one of the great spiritual guides of the twentieth century.

Far too many souls waste time and energy blaming the wrongs in the world on everyone else instead of recognising that they can do something about it when they start within themselves. Start with putting your own house in order first. When a stone is thrown into the centre of a pond, the ripples go out and out; but they start from that stone; they start from that centre. Start with yourself; then you can radiate peace, love, harmony and understanding out to all souls around you. Get into action now. You long to see a better world; then do something about it, not by pointing your finger at everyone else but by looking within, searching your heart, righting your own wrongs and finding the answer within yourself. Then you can move forward with authority and be a real help to your neighbour and to all those souls you contact. Change starts with the individual, then goes out into the community, the town, the nation and the world. (Eileen Caddy, from her book of meditations *Opening Doors Within.*)

Looking into the future, Ecovillages hold out hope.

There is no such thing as waste, only resources looking for a use.

Community Profile

EcoMe

EcoMe is a temporary project to create a safe space for Israelis, Palestinians, Jews, Muslims and Christians to get together and explore their differences and similarities in an environmental context. First established by a group of alumni from the Arava Institute for Environmental Studies, their aims included both ecological awareness and peacework across the political, ethnic and religious fault lines of the Middle East.

It started in the winter 2010 to 2011, and has been set up again each winter since then. It has been functioning as a seasonal project for about four to six months every winter.

EcoMe is a community-building project for Israelis, Palestinians and internationals in the Dead Sea area of the West Bank. It is located just outside of the Palestinian city of Jericho, in Israeli-occupied Area C. The location is accessible to both Palestinians and Israelis, so it is an ideal place to bring people together for retreats and workshops on intentional community-living, sustainability, and nonviolent (compassionate)

Around the fire, a gathering of Israelis and Palestinians.

communication. These programs have been geared to people of all nationalities and faiths.

EcoMe has established itself as a welcome and transformative environment for participants of all backgrounds. No weapons, alcohol, or illegal substances are allowed on the premises, and they maintain a strictly vegetarian kitchen. Accommodation is very basic, and participants benefit from learning together about desert survival and sustainable living.

EcoMe trainings bring together communities of people who are interested in learning and embodying the principles of Nonviolent Communication in their own lives and sharing it with others. Residential multi-day trainings enable greater submersion into the philosophy and practice of NVC and contribute to a deeper experience of community and learning. They provide a rare opportunity to learn from international trainers with wide-ranging experience with other regions and conflicts. They also offer workshops on Permaculture, mud building, and have low technology solar heated showers and composting toilets. Many of the key members are active in the Global Ecovillage Network, and in other environmental and peace projects.

Read more on: http://rebuildingalliance.org/ecome-center/

See also The Arava Institute for Environmental Studies: http://arava.org

Solar panels for the electronics that are needed to stay in touch with the wider society.

12

Appendix 1: Dragon Dreaming: Its power and promise for ecovillages

John Croft and Kosha Joubert

Why is it that so many groups of people planning to create ecovillages fail in their task? It would seem that there is a failure to realise the necessary relationships and connections, participatory processes and co-creative precision and timing. The conventional corporate system of business as usual, built on a worldview that separates us from one another and from nature, tends to disempower individuals, destroy communities and damage the ecosystems upon which we depend. But there are alternatives!

Dragon Dreaming is a project development methodology based upon the principles of personal and group empowerment. Drawing upon the insights and experience of Living Systems Theory, Deep Ecology, Transpersonal Psychology, group dynamics, and organisational development, Dragon Dreaming uncovers the blockages that limit the effectiveness of groups of people to achieve a common task.

Within this framework, any project is seen, primarily, as an opportunity to build a bridge towards the realisation of our dreams, and, in the process, to heal the separation between 'Self' and 'Other' and between 'Theory' and 'Practice'. Any project, to become sustainable, needs to be born from dreams and visions that inspire us deeply, and complete a full cycle of Dreaming – Planning – Doing – Celebrating. Conventional project development focuses on Planning and Doing, activities that take place mainly in the left halves of our brain. We need to balance these with the more intuitive processes of Dreaming and Celebration, which are more predominantly connected to the right halves of our brain, to ensure our projects stay on track and in tune with the needs of the collective.

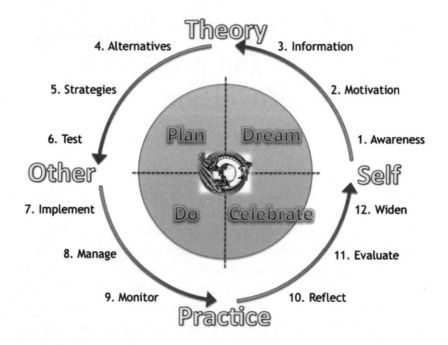

We believe that around 90% of projects get blocked in the Dreaming Stage. This is because the dreams of an individual fail to become the dream of a collective group who could effectively work together to build a project like an ecovillage. Our current educational system does not usually teach us to share those dreams that truly move us with others or to follow our inner compass of inspiration through life. Also, we have not learnt how to share our dreams with others in a way that allows them to take ownership. Many ecovillage projects fail because of what we call the 'founder syndrome'.

And even if we manage to find and build a good team, many groups pass immediately to implement their dreams and the project then fails in a second way: People do not plan to fail, they only fail to plan effectively. Conventional planning separates an elite or 'head' of a project from the 'body' or the 'hands' who are expected to do the work, creating two groups who can blame each other when things go wrong. As a result around 90% of projects fail to work according to plan.

Then as with small business initiatives or community organisations, 90% fail and disband within 3–4 years, because they fail to consider a fourth step. It is Celebration (a process of reflection, evaluation and widening perspectives) that builds the connection from the Planning and Doing stages of a Project back to the collective Dream. From our experience we have found that 25% of the time and resources of a project should ideally be expended on Celebration in order for the project to remain alive and successful into the future.

These statistics show that conventionally only one in a thousand initiatives achieve fruition. People are giving up on their dreams. Aboriginal people say people who lose their Dreaming have lost part of their soul, and will perish quickly. And the symptoms of soul loss are everywhere apparent. Apathy, fatalism, powerlessness, disconnection from both our past ancestors and future descendants, surviving from day-to-day, an endless search for happiness through a culture of consumerism, and chronic depression abound. These are the symptoms of soul loss associated with the loss of a belief that Dreams can come true!

Dragon Dreaming as a training methodology is now being applied in more than 2,950 projects in more than thirty-seven countries around the world, from Brazil to Russia and from Canada to the Congo, at community level, within government organisations, in businesses and with training organisations and universities. By working within more healthy relationships and by allowing close feedback loops to generate continuous learning and adaptation, we generate new degrees of commitment and satisfaction, so our intentions are translated into new commitment to ensure our goals are reached.

Dragon Dreaming, through replacing conventional win-lose or lose-lose games with a win-win-win approach, opens up unpredictable possibilities. Using the Dream-Plan-Do-Celebrate Wheel, projects are a win for the people involved in fostering personal growth, a win for the communities in which they live, and a win for the planet as a whole in their commitment to work in service to the Earth, for the wellbeing and flourishing of life itself. Bringing groups of people together around such a shared intention is a very powerful tool.

Using a participatory strategic planning method pioneered through working with Australian Aboriginal groups, in Papua New Guinea and around the Third World, Dragon Dreaming creates a Karabirrdt Critical Path Diagram, a game in which players come together to achieve their dreams.

Dragon Dreaming is four things: It is a Dream of building a new win-win-win culture for the Great Turning of our culture away from suicidal culture built on destruction of its own life support systems to one that sustains complex life on the planet. It is a Planning Philosophy for the design of successful projects. It is a Toolbox of techniques that enable people to do what they dream about, and it is a Community of Practitioners, who through Celebration, support each other in making dreams come true.

For more information, see: www.dragondreaming.org. There you will find trainers who can assist, a list of forthcoming events, resource materials, and a free open source ebook to assist you get started in making your dreams for an ecovillage come true.

Appendix 2: The Ecovillage Design Mandala

Kosha Joubert and Robin Alfred

Kartong is a village in The Gambia, closely related to an Eco-Retreat and Learning Centre, called Sandele. In 2013, after a meeting with leaders of GEN, the Kartong village elders decided that they wished for Kartong to transition to an ecovillage. The EDE was seen as the perfect first step. One year later, with some funding secured largely from a German Foundation, 25 participants from Kartong were carefully selected by an official village panel to ensure representation and diversity. Together, and with the full backing of the village elders, these 25 embarked on a journey through the mandala of the EDE. To be honest, it is hard to imagine how a traditional village can make a transition from where it is now without the groundwork of inspiring a critical mass of people and giving them the space to dream, plan, do and celebrate their own pathway into the future using EDE principles.

The first day of the Ecovillage Design Education (EDE) in Sandele, Gambia, turns into a celebration as the villagers of Kartong arrive, bringing with them their songs, their dances and their expert story-telling abilities. We receive our first learning immediately: whenever one member of the community breaks into a new song (often created on the spot), the rest of the group support whole-heartedly, finding the rhythm, the melody and harmonies in a generous outflowing of untethered encouragement. Just imagine if we had all received such a joyful reception of our creativity from our communities – how much easier it would be to muster the courage to express!

In 2004, Gaia Education was born as the educational arm of the

Global Ecovillage Network (GEN). The idea of bringing the learning of ecovillages out to inspire a broader public had been around for a while. Now it was time to realise the dream. In May 2004, 24 ecovillage educators, amongst them Kosha, met in beautiful Findhorn, Scotland to bring together the best of what they knew and found precious to share. The Ecovillage Design Education (EDE) Curriculum was born and became an official contribution to the UN Decade for the Education for Sustainable Development, 2004–2014.

The hard part was not to bring out the immense wealth of expertise from the ecovillage networks, but to boil this down to a core-pattern that was intricate enough to be a map of reality, while at the same time simple enough to speak to all parts of a movement as diverse as this one.

The Ecovillage Design Mandala picture

The Ecovillage Design Mandala achieved this marriage of simplicity and complexity, being a design tool that, if followed, helps to ensure a holistic path to sustainable development, embracing the *social, worldview, ecological* and *economic* dimensions of sustainability.

The mandala can be applied to systems on all levels: to an organisation, intentional community, traditional village, urban neighbourhood, region, etc.

GEN currently uses a form of the EDE-mandala that adds *integral or participatory design* as a fifth dimension and puts it at the centre of the wheel. Also, the word *'worldview'* is often replaced by *'culture'*. In all cases, the EDE course touches on all aspects, so that the mandala becomes a map that enables us to explore the territory of sustainability together.

Using the EDE-mandala for a four-week process of design has several benefits:

In this participatory process, the participants themselves *identify the strengths and weaknesses of their system/village* in each of the dimensions of the mandala. How connected are individuals to nature? How strong is a sense of self-worth and cultural celebration? How are conflicts dealt with? How is leadership shared? How are natural resources used and eco-systems regenerated? How sustainable are people economically? These, and many more questions are with us as we probe and explore, while also allowing ourselves to be inspired by best-practice examples from all over the world.

The participants then go on to *identify and design leverage points* in each of the dimensions: steps that can be taken, projects that can be realised, that will have a maximum impact for minimum effort. We start by harvesting the low-hanging fruits. An experience of success helps maintain momentum!

Within the group, we have a chance to find out *who naturally takes responsibility for which parts of the emerging design,* while at the same time *giving everyone a sense of ownership* over the whole process. Teams emerge that refine the designs for particular areas. These teams draw in others from the community to take implementation forward after the course. If any tensions arose these provided welcome input for teaching conflict facilitation skills.

Through the depth of the process of spending four weeks together, sharing and learning, witnessing and supporting each other's vulnerability and courage, a strong *network of mutual recognition and trust* is built amongst participants which carries on into the future. Regular meetings take place with all stakeholders. We take care in every EDE to invite some participants from foreign countries as these cross-cultural contacts prove to be of immeasurable importance.

The two hoped for outcomes of the Kartong EDE were:

1. To provide proposals and designs for how the village can move towards becoming an ecovillage, and
2. To develop a plan for the tourism zone in the Kartong area.

The village stakeholders were consulted during the process, so that the design process took place in agreement with them. The Director of The Gambia Tourism Board congratulated the participants on their hard work and stated that he had, 'No choice but to accept the proposal for the Tourist Development Area'.

Today, the EDE has been delivered in over thirty-four countries on six continents. Each course has birthed a multitude of projects. While the curriculum is designed to be adapted to the local context as needed, a natural flow can be to start with the *cultural* or *worldview* dimension, giving space to a process of individual reconnection to self, community, inner purpose and nature. Design groups are formed that have interest in particular project areas. Then the time may be ripe to dive into the *social,* looking at the skills that are needed for the whole of a community to become more than the sum of its parts. All learning is integrated into the designs – so that the social aspect is always taken into account.

From there, a deeper exploration of *ecology* and *world* becomes held within a resilient community context, including applying learning about sustainable food production, energy generation, building techniques, water harvesting and waste water treatment and recycling processes. All these topics are woven into the designs. Finally, we move to *economy* and a look at how to create economically thriving communities, be it through providing ethical investment opportunities, creating a local currency or trading system, generating employment and social enterprise, or by understanding the potential of communities to provide positive alternatives to the malfunctioning mainstream economic system. For the actual project designs, this is the phase of creating a business plan – making sure the vision and design are economically viable and sustainable and will be implemented.

The ever-evolving EDE is now also available for online study and has generated four accompanying books – one for each of the four main dimensions of sustainability it embraces (see Further Reading).

Appendix 3: Transition to Resilience

Robin Alfred and Kosha Joubert

'The Transition to Resilience (T2R) Training – A Learning Adventure for Change Makers', was born in 2012 out of a desire to embed the design and implementation of sustainability and resilience in communities, regions and workplaces in a deep and ongoing learning process.

Existing trainings, though excellent, were run as one-off training events, be they a two-day Transition Training, a two-week Permaculture Design training, or a one-month Ecovillage Design Education. They are all terrific programmes, yet we found ourselves consistently meeting former participants some months after they had completed a training, saying, "Wow! It's much harder than I thought to actually implement what I have learned..." Change makers and their projects needed further support to sustain momentum and motivation.

As we start putting into practice the inspiration and skills gained, we inevitably run into the resistance of systems to change. We need to deepen our understanding about the process of change over time, and learn how to be personally resilient within it, becoming able to embrace challenges and failures alongside the successes. We need to learn to build and sustain teams, and to continue to adapt what we are doing to meet the ever-changing demands and needs of our environments.

The T2R was designed to meet these needs and to help people ground their visions and designs in real projects that make a difference in their regions, workplaces and communities.

Hatched out of two-year European Union-funded collaboration between eight existing training providers and change agencies (GEN,

Transition Town Network, Permaculture Academy, The Hub Network, and the Ecovillages of Findhorn Foundation, Cloughjordan, Damanhur and Sieben Linden) we quickly saw that the heart of the learning process was not accruing more knowledge about technologies and *content*, but understanding the *process* of how change happens and what it takes to be a change maker.

Participants can more and more easily access information and knowledge from the Internet, particularly with regard to sustainable technologies such as building techniques, energy provision, food production and recycling (please also see the GEN Solution Library - www.solution.ecovillage.org). What is not so easy to come by is a deeper understanding of the nature of change and the capacity to develop both personal and group resilience to enable us to effectively meet the needs of the world.

In T2R we decided, therefore, to work at three levels, I / We / World, in order to develop a holistic and sustainable approach to processes of change.

At the 'I' Level, we work to help people strengthen their personal resilience through developing an integral practice that will sustain them physically, emotionally, mentally and spiritually. Too often we forget that sustainability starts with our own person and that we cannot build sustainable teams and systems while landing in burnout

ourselves. 'Being the change we want to see' also entails ensuring that our own lifestyles and relational networks are such that we can live fulfilled, healthy lives.

The 'We' level focuses on the resilience of groups and teams, using both the training group itself and the groups that participants are initiating or may already be a part of as fields of practice. How is the group dynamic? How do power, rank and status operate? What is the group's vision and purpose? How does collective wisdom arise in the group? How do we ensure that we keep our circles open as we build teams and projects? Keeping these, and many other inquiries and questions alive, helps to maintain a healthy group life while enabling the whole to become significantly greater than the sum of its parts, even when the going gets rocky.

And, of course, we then need to relate the 'I' and the 'We' to the needs of the world. To be aware of both global and local needs and challenges, be they climate change and the global economy, or the need for local car-pooling or a more effective recycling system. How is what we are doing, and what I am passionate about, meeting the needs of our local and global communities? How do we adapt our designs to respond to changes in the world?

As we delved into the archetypal nature that learning journeys for change makers take, we found that the map of the 'Hero's Journey', symbolising a very ancient pattern of human transformation, first identified by US mythologist Joseph Campbell and outlined in his book, *The Hero With a Thousand Faces,* was of great relevance. Quickly seeing that this template matched, as an archetypal map would, the stages of the learning journey that we were mapping, we embedded our curriculum within its core elements.

The journey of transformation and change combines both masculine and feminine phases, times when we need to bring out our inner masculine to set clear boundaries or follow-through, and times when we need to bring out our inner feminine to hear the call, embrace the diversity of a fellowship, or go deep within in order to cross the dark night of the soul. A deeper understanding of the complexity of whole systems change processes equips us with a helpful map as we set out to traverse new territories.

We decided to develop the curriculum of the T2R Training as a pattern language with core concepts and questions being addressed in outline only, with directions and references offered, thus leaving space for local trainers to add their particular colour and flavour in the detailed implementation. In addition, we saw the need for study and support between modules and so the T2R learning template envisions people working in action learning sets in between modules.

At the time of writing, mid-2014, the first T2R training is being completed at the Findhorn Foundation: an eighteen-month learning journey spanning 4 x 7-day and 1 x 5-day modules, and an application has gone into the EU for a 3-year partnership programme which would see the curriculum deepened and spread across Europe through collaboration with universities, social enterprises and other training providers.

Follow up and Further Reading

General

Communities Magazine
http://www.ic.org/communities-magazine-home/

Guardian Weekly
http://www.theguardian.com/weekly

Permaculture Magazine
http://www.permaculture.co.uk

Resurgence and Ecologist Magazine
http://www.resurgence.org/magazine/resurgence-ecologist.html

The 4 Keys:

Dawson, Jonathan; Jackson, Ross and Norberg-Hodge, Helena, eds. 2006. *Gaian Economics – Living well within Planetary Limits*. Permanent Publications.

Alfred, Robin and Joubert, Kosha, eds. 2007. *Beyond You and Me – Inspiration and Wisdom for Building Community*. Permanent Publications.

Lindegger, Max and Mare, Chris, eds. 2009. *Designing Ecological Habitats – Creating a Sense of Place*. Permanent Publications.

Harland, Maddy and Keepin, Will, eds. 2012. *The Song of the Earth*. Permanent Publications.

Chapter 1. Ecovillages

Allen, Reginald, ed. 1966. *Greek Philosophy: Thales to Aristotle*. The Free Press, UK.

Ansell, Vera; Coates, Chris; Dawling, Pam; How, Jonathan; Morris, William and Wood, Andy eds. 1989. *Diggers and Dreamers*. Communes Network Publication, England.

Christensen, Karen and Levinson, David. 2003. *Encyclopaedia of community*. Sage Publications, California, London and New Delhi. Four volumes.

Christian, Diana Leafe. 2003. *Creating a Life Together*. New Society Publishers, Canada.

Communities Directory, 2005 Edition. Fellowship for Intentional Community, USA.

Ecovillage Education:
http://www.gaiaeducation.net/

Eurotopia:
http://www.eurotopia.de/wvk-e-resource.html

Findhorn footprint:
http://www.ecovillagefindhorn.com/docs/FF%20Footprint.pdf

Gering, Ralf. 2009. *Encyclopedia of Large Intentional Communities*. Only available online: www.comuntierra.org/site/downloads.php?id=9&id_idioma=2

Gilman, Diane and Robert. 1992. *Eco-villages*. In Context Institute, Bainbridge Island, WA, USA.

Gilman, Robert. September 2008:
http://www.ecovillagenews.org/wiki/index.php/Robert_Gilman_on'Multiple_centers_of_Initiative'

Global Ecovillage Network (GEN) – Europe. 1998. *Directory of Ecovillages in Europe*.

Global Ecovillage Network:
http://gen.ecovillage.org

Jackson, Hildur and Svensson, Karen. (eds). 2002. *Ecovillage Living*. Green Books and Gaia Trust, UK.

Kozeny, Geoph. Film:
http://www.ic.org/from-visions-of-utopia-to-the-many-faces-of-community/

Low impact living communities in Britain. 2014. A *Diggers and Dreamers* Review.

New Lanark:
http://en.wikipedia.org/wiki/New_Lanark

Owen, Robert:
http://www.robert-owen.com/

Philo. All quotes taken from:
http://www.jewishencyclopedia.com/articles/5867-essenes

Plato. All quotes taken from:
Plato. 1955. *The Republic.* Penguin Classics, UK.

Pythagoras. Source of information:
http://www.thefullwiki.org/Pythagoras
Sargent, Lyman Tower. 2010. *Utopianism.* Oxford University Press, UK.

Chapter 2. Permaculture Tools

Bell, Graham. 1992. *The Permaculture Way.* Thorsons, UK.

Biomimicry:
http://biomimicry.net/about/
Carson, Rachel. 1965. *Silent Spring.* Penguin, UK.
Chamberlin, Shaun. 2009. *The Transition Timeline.* Green Books, UK.
Hawken, Paul. 2008. *Blessed Unrest.* Penguin, UK.
Holmgren, David and Mollison, Bill. 1978. *Permaculture One.* Transworld Publishers, Australia.
Holmgren, David. 2002. *Permaculture. Principles and Pathways Beyond Sustainability.* Holmgren Design Services, Australia.
Mollison, Bill and Slay, Reny Mia. 1991. *Introduction to Permaculture.* Australia. Tagari Publications.
Mollison, Bill. 1988. *Permaculture – A Designers' Manual.* Tagari Publications, Australia.

Transition Towns:
http://www.transitionnetwork.org/news

Chapter 3. Size of Communities

'A New We', a film by L.O.V.E. Productions, 2010:
http://www.newwe.info
Alexander, Christopher. 1977. *A Pattern Language.* Oxford University Press, USA.
Allen, Joan de Ris. 1990. *Living Buildings.* Camphill Architects, UK.
Bettelheim, Bruno. 1971. *The Children of the Dream.* Granada Publishing, Great Britain.

Co-housing websites:
http://www.cohousing.org
http://co-housing.co.uk

Chapter 4. Farming

Biodynamic farming organisations:
https://www.biodynamics.com

Community Supported Agriculture:
http://www.nal.usda.gov/afsic/pubs/csa/csa.shtml
Groh, Trauger and McFadden, Steven. 1997. *Farms of Tomorrow Revisited.* Biodynamic
    Farming and Gardening Association, USA.
Hart, Robert. 1996. *Forest Gardening.* Green Earth Books, UK.

Integrated Pest Management:
http://www.epa.gov/opp00001/factsheets/ipm.htm
Koepf, Herbert and Pettersson, Bo and Schaumann, Wolfgang. 1976. *Bio-dynamic
    Agriculture.* Anthroposophic Press, USA.

La Via Campesina:
http://viacampesina.org/en/
New View magazine:
http://www.newview.org.uk/new_view.htm

Organic farming organisations:
http://guides.lib.ndsu.nodak.edu/content.php?pid=324631&sid=2656969
United Nations Environment Programme's report 'Towards a Green Economy', 2011:
http://www.unep.org/greeneconomy/GreenEconomyReport/tabid/29846/Default.
    aspx.

The United Nations Food and Agriculture Association:
http://www.fao.org/publications/card/en/c/33a0aa55-7438-48ea-a5e3-
    6f767acb217b/

Urban farming references:
Sheffield, UK: http://growsheffield.com/abundance/
Local Greens in London, UK: http://www.localgreens.org.uk/
Eagle Street Rooftop Farm, New York City: http://rooftopfarms.org/
Bright Farms, New York City: http://brightfarms.com/s/#!/our_farms

Chapter 5. Gardening, Soil and Plants

Allen, Jenny. 2002. *Smart Permaculture Design.* New Holland, Australia.
Ball, D., and Davis, B., and Fitter, A., and Walker, N. 1992. *The Soil.* Harper Collins
    Publishers, UK.

Bates, Albert. 2010. *The Biochar Solution: Carbon Farming and Climate Change.* New Society Publishers, USA.

Jenkins, Joseph. 2012. The *Humanure Handbook: A Guide to Composting Human Manure.* Joseph Jenkins, Inc. USA.

Living Routes:
http://www.environment-ecology.com/ecovillages/422-living-routes-study-abroad-in-ecovillages.pdf

Soil triangle:
http://en.wikipedia.org/wiki/Soil_texture

Chapter 6. House Design

Active House:
http://www.activehouse.info/about-active-house/specification
Alexander, Christopher. 1977. *A Pattern Language.* New York. Oxford University Press.

Passive House:
http://www.passiv.de/en/index.php
Day, Christopher. 1990. *Places of the soul.* Aquarian/Thorsons, UK.
Pearson, David. 1994. *Earth to Spirit.* Gaia Books, UK.

Chapter 7. Building

Bainbridge, David; Steen, Athena and Steen, Bill. 1994. *The Straw Bale House.* Chelsea Green Publishing Company, USA.
CINVA Ram:
http://www.dirtcheapbuilder.com/Home_Building/Earth_Block_Construction.htm

Earthships:
http://en.wikipedia.org/wiki/Earthship
Falu Paint: http://falurodfarg.com/eng/falun-red-paint/the-original/
Pearson, David. 1989. *The Natural House Book.* Gaia Books, UK.

Chapter 8. Energy and Technology

Biofuels:
http://www.euractiv.com/climate-environment/biodiesels-pollute-crude-oil-leaked-
    eu-data-news-510437
The Farm Ecovillage:
http://www.thefarm.org/lifestyle/albertbates/akb.html

Chapter 9. Water

Etnier, Carl and Guterstam, Bjørn. Eds. 1991. *Ecological Engineering for Wastewater
    Treatment.* Bokskogen, Gothenburg, Sweden.
Schwenk, Theodor. 1996. *Sensitive Chaos.* Rudolf Steiner Press, London.
Wilkes, John. 2003. *Flowforms.* Floris Books, Edinburgh.
Yoemans, P. A. 1981. *Water for every farm/Using the keyline plan.* Second Back Row
    Press, Australia.

Chapter 10. Alternative Economics

Citizen's wage:
http://www.citizensincome.org/
Douthwaite, Richard. 1996. *Short Circuit.* Green Books and Lilliput Press, Ireland.
Douthwaite, Richard. 1999. *The Ecology of Money.* Green Books, UK.

Ethical banks:
http://www.betterworldclub.org/?p=2663

Gross National Happiness:
http://www.grossnationalhappiness.com/
Hayter, Teresa. 1981. *The Creation of World Poverty.* Pluto Press, UK.
Henderson, Hazel. 1978. *Creating Alternative Futures.* Perigee Books, USA.
Hopkins, Rob. 2013. *The Power of Just Doing Stuff.* Transition Books, UK.
Kennedy, Margrit. 1995. *Interest and Inflation Free Money.* Seva International, USA.

Natural Economics:
www.globalcanopy.org
Sahlins, Marshall. 1974. *Stone Age Economics.* Tavistock Publications, UK.
Schumacher, E. F. 1975. *Small Is Beautiful.* Harper and Row, USA.
Wright, David H. 1979. *Cooperatives and Community.* Bedford Square Press, UK.

Chapter 11. How Groups Develop

Jackson, Hildur ed. 1999. *Creating Harmony*. Gaia Trust, Denmark.
Macy, Joanna and Johnstone, Chris. 2012. *Active Hope: How to Face the Mess We're in Without Going Crazy*. New World Library, USA.

Open Space meeting processes:
http://www.openspaceworld.org/
Langford, Andy. *Designing Productive Meetings and Events*. Available from Jane Hera:
     www.marvellousmeetings.co.uk

Chapter 12. Crises and Solutions

Ecological Footprint:
www.footprintstandards.org
http://en.wikipedia.org/wiki/Ecological_footprint - cite_note-7#cite_note-7
Caddy, Eileen. 1987. *Opening Doors Within*. Findhorn Press, UK.

Fracking:
www.keeptapwatersafe.org/global-bans-on-fracking/

Fuller, R. Buckminster. 1969. *Operating manual for Spaceship Earth*. Southern Illinois University Press, USA.
Holdredge, Craig. 1996. *Genetics and the Manipulation of Life*. Lindisfarne Press, UK.
Koestler, Arthur. 1961. *The Yogi and the Commissar*. Collier Books. New York.
Pachamama Alliance Symposium:
http://www.awakenthedreamer.com/
Seamon, David and Zajonc, Arthur, eds. 1998. *Goethe's Way of Science*. State University of New York Press, USA.
Suchantke, Andreas. 2001. *Eco-Geography*. Floris Books, Edinburgh.
Rees, William and Wackernagel, Mathis. 1996. *Our Ecological Footprint*. Canada. New Society Publishers, Canada.
Wilkes, John. 2003. *Flowforms*. Floris Books, Edinburgh.
Zoeteman, Kees. 1991. *Gaia-Sophia*. Floris Books, Edinburgh.

Index